Lecture Notes in Logic 3

William J. Mitchell   John R. Steel

Fine Structure and Iteration Trees

Springer-Verlag
Berlin Heidelberg New York
London Paris Tokyo
Hong Kong Barcelona
Budapest

Authors

William J. Mitchell
Department of Mathematics
University of Florida
Gainesville, FL 32611-2082, USA
E-mail: mitchell@math.ufl.edu

John R. Steel
Department of Mathematics
University of California, Los Angeles
Los Angeles, CA 90024-1555, USA
E-mail: steel@math.ucla.edu

Mathematics Subject Classification (1991): 03E45, 03E55, 03E10, 04A10, 04A15

ISBN 3-540-57494-8 Springer-Verlag Berlin Heidelberg New York
ISBN 0-387-57494-8 Springer-Verlag New York Berlin Heidelberg

Printed in Germany

SPIN: 10061642 46/3140-543210 - Printed on acid-free paper

Contents

§0. Introduction

In these notes we construct an inner model with a Woodin cardinal, and develop fine structure theory for this model. Our model is of the form $L[\vec{E}]$, where $\vec{E}$ is a coherent sequence of extenders, and our work builds upon the existing theory of such models. In particular, we rely upon the fine structure theory of $L[\vec{E}]$ models with strong cardinals, which is due to Jensen, Solovay, Dodd-Jensen, and Mitchell, and upon the theory of iteration trees and "backgrounded" $L[\vec{E}]$ models with Woodin cardinals, which is due to Martin and Steel. Our work is what results when fine structure meets iteration trees.

One of our motivations was the desire to remove the severe limitations on the theory developed in [MS] caused by its use of an external comparison process. Because of this defect, the internal theory of the model $L[\vec{E}]$ constructed in [MS] is to a large extent a mystery. For example it is open whether the $L[\vec{E}]$ of [MS] satisfies GCH. Moreover, the use of an external comparison process blocks the natural generalization to models with infinitely many Woodin cardinals of even the result [MS] does prove about $L[\vec{E}]$, that $L[\vec{E}] \models \mathrm{CH} + \mathbb{R}$ has a definable wellordering.

Our strategy for making the comparison process internal is due to Mitchell and actually predates [MS]. The strategy includes taking finely calibrated partial ultrapowers ("dropping to a mouse") at certain stages in the comparison process. Thus to define the internal comparison process and prove it succeeds one needs fine structure theory. Of course, fine structure theory requires a comparison process, but fortunately we are led not into a vicious circle, but into a benign helix: that is, an induction. The whole of what follows can be viewed as a long inductive proof that a certain construction yields a model $L[\vec{E}]$ whose levels have certain fine structural properties. Among those properties is a strong local form of GCH.

We have as a corollary that if ZFC + "There is a Woodin cardinal" is consistent, then so is ZFC + "There is a Woodin cardinal" + GCH. But our interest is not so much in this relative consistency result, which can probably be proved more easily using forcing, as in the inner model $L[\vec{E}]$ itself, and the fine structure techniques which should eventually decide many questions about $L[\vec{E}]$ and similar models containing more Woodin cardinals.

The model $L[\vec{E}]$ and its fine structure theory are likely indispensable for proving certain relative consistency statements in which the theory hypothesized consistent does not directly assert the existence of large cardinals. For example the following conjecture is widely believed to be true:

Conjecture. *If ZFC + "There is an ω_2-saturated ideal on ω_1" is consistent, so is ZFC + "there is a Woodin cardinal".*

Of course, the conjecture is really that the relative consistency is provable in Peano Arithmetic. Shelah has proven the converse relative consistency result. Mitchell ([M?]) has proved the conjecture with its conclusion weakened to "ZFC + $\exists\kappa(o(\kappa) = \kappa^{++})$ is consistent". The present paper is a step toward extending Mitchell's arguments so as to prove the full conjecture. What we lack at the moment is a method which does not use large cardinals in V for showing that a certain $L[\vec{E}]$ type model is sufficiently iterable. This "Core model iterability" problem is one of the key open problems in the area. Its solution should lead to a proof of the conjecture, and to much more.

The notes are organized as follows. In §1 we introduce potential premice, which are structures having some of the first order properties of the levels of the model we eventually construct. Perhaps the most notable thing here is that the extender sequence $\vec{E}^{\mathcal{M}}$ of a potential premouse (ppm) $\mathcal{M}$ may contain extenders which do not measure all sets in $\mathcal{M}$. In general, an E on $\vec{E}^{\mathcal{M}}$ measures only subsets of crit E constructed in $\mathcal{M}$ before the stage at which E was added to $\vec{E}^{\mathcal{M}}$. This tactic, which is due to S. Baldwin and Mitchell, greatly simplifies fine structure theory.

Section §2 studies definability over potential premice. We introduce the $r\Sigma_n$ hierarchy, a slight variant on the usual Levy hierarchy. We follow Magidor and Silver in introducing Skolem terms so as to avoid proving $r\Sigma_n$ uniformization, and in working directly with $r\Sigma_n$ formulae rather than master codes and iterated $r\Sigma_1$ definability. We show that being a ppm is preserved under the appropriate embeddings. Finally, we introduce projecta, standard parameters, solidity and universality of parameters, cores, and soundness. These are standard fine structural notions, with the exception of solidity.

The analysis of §2 is not appropriate for a certain sort of ppm, the "active type III" variety. In §3 we modify it slightly so that it suits these ppm. This leads to an annoying case split in the details of many arguments, a split which we have sometimes ignored.

One important feature of the Baldwin-Mitchell tactic is that all levels of the model we build will be completely sound. Ultrapowers of sound structures can be unsound, but all proper initial segments of the ultrapower will be sound. So it suffices to consider only ppm all of whose proper initial segments are sound. These we call premice.

In §4 we define the $r\Sigma_n$ ultrapower $\mathrm{Ult}_n(\mathcal{M}, E)$ of a ppm $\mathcal{M}$ by an extender E measuring all sets in $\mathcal{M}$ and satisfying crit $E < \rho_n^{\mathcal{M}}$. We prove Los' theorem and show that the canonical embedding is $r\Sigma_{n+1}$ elementary if $\mathcal{M}$ is n-sound. We show that if $\rho_{n+1}^{\mathcal{M}} \leq$ crit E, $\mathcal{M}$ is n-sound, and E is "close to being a member of $\mathcal{M}$", then the canonical embedding preserves the $n + 1^{\text{st}}$ standard parameter, *provided this parameter is solid.* This result explains the importance of solidity.

Section §5 introduces iteration trees and n-iterability. It also proves the Dodd-

Jensen lemma on the minimality of iteration maps, which is a key tool in our work.

In Section §6 we investigate the uniqueness of wellfounded branches in iteration trees. Theorem 6.1 is a straightforward generalization of the uniqueness theorem of [MS]. Theorem 6.2 is a fine structural strengthening of theorem 6.1 which takes considerably more work to prove. Theorem 6.2 has the important consequence that all the iteration trees we care about have at most one cofinal wellfounded branch.

Section §7 proves a comparison lemma for iterable premice. The lemma is never used in what follows, but the method of proof, the comparison process, is used throughout.

In §8 we prove our main fine-structural result: the $n+1^{st}$ standard parameter of an n-sound, n-iterable premouse is $n+1$-solid and $n+1$-universal. The method of proof traces back to Dodd's proof that GCH holds in the model of [D]. The method also gives a useful condensation result, Theorem 8.2.

In §11 we finally construct (assuming there is a Woodin cardinal in V) some iterable premice. We in fact construct a model $L[\vec{E}]$ with a Woodin cardinal all of whose levels are ω-sound and ω-iterable premice. §9 and §10 are devoted to some preliminary lemmas which guarantee that the construction of §11 puts enough extenders on $\vec{E}$ that we do indeed get a model with a Woodin cardinal. Section §12 shows that the construction of §11 produces an iterable structure $L[\vec{E}]$ by associating to any iteration tree on $L[\vec{E}]$ an iteration tree on V and then using the results of [MS].

We did the work described here during 1987–1989 and wrote it up in a set of notes which has been informally circulated since October 1989. This paper is essentially identical to that set of notes. We wish to thank Kai Hauser, Mitch Rudominer and Ernest Schimmerling for reading those notes carefully and bringing errors to our attention.

Since 1989 the theory described here has advanced in several ways. In the spring of 1990, Steel found a solution to the core model iterability problem mentioned above, and with it was able to extend the work of [M?] to the level of a Woodin Cardinal [S?a]. He used this to show that if there is a saturated ideal on ω_1, together with a measurable cardinal, then there is an inner model with a Woodin cardinal. The measurable cardinal should not be necessary here and its use may indicate a weakness in the basic theory of [S?a]. As expected, other relative consistency results have come out of this work. Some of these use the weak covering lemma for the model of [S?a], which was proved in late 1990 by Mitchell [MSS?].

Schimmerling [Sch] has investigated the combinatorial set theory of the model $L[\vec{E}]$ described in this paper. He showed that $\Box_{\omega_1}$ holds in this model, and that weak $\Box_\kappa$ holds for all κ. It is open whether $L[\vec{E}]$ satisfies $\forall\kappa\,\Box_\kappa$. Schimmerling was able to combine his work on $\square_\kappa$ with the ideas of [S?a], [MSS?] and arguments of Todorcevik and Magidor in order to show that the proper forcing axiom implies that there is an inner model with a Woodin cardinal.

Finally, Steel ([S?b], [S?c]) has extended the theory presented here to models having more Woodin cardinals.

§1. Good Extender Sequences

DEFINITION 1.0.1. Let $\kappa < \nu$ and suppose that M is transitive and rudimentarily closed. We call E a (κ,ν)-*extender over M* iff there is a nontrivial Σ_0-elementary embedding $j : M \to N$, with N transitive and rudimentarily closed, such that $\mathrm{crit}(j) = \kappa$, $j(\kappa) > \nu$, and

$$E = \{(a,x) \mid a \in [\nu]^{<\omega} \wedge x \subseteq [\kappa]^{\operatorname{card} a} \wedge x \in M \wedge a \in j(x)\}\,.$$

We write $\kappa = \operatorname{crit} E$, $\nu = \operatorname{lh} E$.

Remark. If the requirement that N be transitive is weakened to $\nu+1 \subseteq \mathrm{wfp}(N)$, where $\mathrm{wfp}(N)$ is the wellfounded part of N, then we call E a (κ,ν) *pre-extender over M.*

We are interested in this weakening for the purely technical reason that if $\nu \leq \mathrm{OR}^M$, then pre-extenderhood is expressible by a simple first order sentence about $(M, \in, E)$.

DEFINITION 1.0.2. Let $S \subseteq \mathrm{OR}$ and suppose $\vec{E} = \langle E_\alpha \mid \alpha \in S\rangle$ is a sequence of extenders (E_α over some M_α). Then

$$J_\alpha^{\vec{E}} = J_\alpha^A$$

where

$$A = \{(\alpha, a, x) \mid \alpha \in S \wedge (a,x) \in E_\alpha\}\,.$$

Remark. So $J_\alpha^{\vec{E}} = J_\alpha^{\vec{E}\restriction\alpha}$. The ordinals in S are the stages at which extenders are activated.

Let $\vec{E} = \langle E_\alpha \mid \alpha \in S\rangle$ be given. Let $\alpha \in S$ and E_α be a (κ,γ) pre-extender. We put

$$\begin{aligned}(a,b,\delta) \in \tilde{E}_\alpha \quad \text{iff} \quad & [\delta < \gamma \wedge b \quad \text{is a function} \\ & \wedge \operatorname{dom} b = \kappa \wedge b \in J_\alpha^{\vec{E}\restriction\alpha} \\ & \wedge a = E_\alpha \cap ([\delta]^{<\omega} \times \operatorname{ran} b)]\,.\end{aligned}$$

We then define

$$\mathcal{J}_\alpha^{\vec{E}} = \begin{cases} (J_\alpha^{\vec{E}}, \in, \vec{E}\restriction\alpha) & \text{if} \quad \alpha \notin S \\ (J_\alpha^{\vec{E}}, \in, \vec{E}\restriction\alpha, \tilde{E}_\alpha) & \text{if} \quad \alpha \in S\,. \end{cases}$$

Remarks. (a) In the sequences $\vec{E}$ we shall consider, $\operatorname{lh}\ E_\alpha = \alpha$ for all $\alpha \in S$.

(b) Of course if $\alpha \in S$ then the sets first order definable over $(J_\alpha^{\vec{E}}, \in, \vec{E}\restriction\alpha, E_\alpha)$ are the same as those first order over $\mathcal{J}_\alpha^{\vec{E}}$, but the latter structure is a better starting point for fine structure.

(c) We are interested only in the case that each E_α, $(\alpha \in S)$ is an extender over $J_\alpha^{\vec{E}\restriction\alpha}$. Thus E_α is to measure no sets constructed after it is activated, a very useful idea due to Baldwin and Mitchell. As a consequence, the subsets of $J_\beta^{\vec{E}}$ in $J_{\beta+1}^{\vec{E}}$ are just those first order definable over $\mathcal{J}_\beta^{\vec{E}}$. In earlier setups one needed also "measure quantifiers" coming from the E_α, $\alpha \leq \beta$, to define the new sets. This complicated the fine structure substantially.

One consequence of the Baldwin-Mitchell idea is that we can work entirely with structures which are strongly acceptable in the sense of Dodd and Jensen [DJ1].

DEFINITION 1.0.3. A structure $(J_\alpha^A, \in, \ldots)$ is *strongly acceptable* iff whenever $\beta < \alpha$ and

$$P(\kappa) \cap (J_{\beta+1}^A - J_\beta^A) \neq \varnothing$$

then $J_{\beta+1}^A \models' \text{card}(J_\beta^A) \leq \kappa$.

Notice that if J_α^A is strongly acceptable and $J_\alpha^A \models$ "κ^+ exists" then $J_\alpha^A \models$ "$P(\kappa)$ exists and $P(\kappa) \subseteq J_{\kappa^+}^A$". In particular, GCH holds in strongly acceptable structures.

It is a basic fact in the fine structure of L that $(J_\alpha^\varnothing, \in)$ is always strongly acceptable. On the other hand, in the usual stratification of $L[\mu]$, $J_{\kappa+2}^\mu$ is not strongly acceptable (for $\kappa = \text{crit}\ \mu$).

Let E be a (κ, ν) extender over M. For $\kappa \leq \xi < \nu$, we say ξ is a *generator of* E iff whenever $a \in [\xi]^n$ and $f \in M$ and $f : [\kappa]^n \to \kappa$, $\xi \neq [a, f]_E^M$ (that is, $\{\langle u_1 \cdots u_n, u_{n+1}\rangle \mid f(u_1 \cdots u_n) = u_{n+1}\} \notin E_{a\cup\{\xi\}}$). (Equivalently, ξ is a generator of E iff ξ is the critical point of the natural embedding from $\text{Ult}(M, E \restriction \xi)$ into $\text{Ult}(M, E)$.)

Thus κ is the least generator of E. All other generators are strictly greater than $(\kappa^+)^M$. Note that the property of being a generator of E depends only on E and $P(\kappa)^M$, and E determines $P(\kappa)^M$.

Let η be the larger of $(\kappa^+)^M$ and $\sup\{\xi + 1 : \xi \text{ is a generator of } E\}$. Then $\eta \leq \nu$, since $M \models \kappa^+$ exists in the models of interest. We call η the *natural length* of E. Suppose $\text{Ult}(M, E)$ is wellfounded, where the ultrapower is formed using functions in M. Regarding $\text{Ult}(M, E)$ as transitive, let $i \colon M \to \text{Ult}(M, E)$ be the canonical embedding. Then $\text{Ult}(M, E) \models$ "η^+ exists" since $\eta < i(\kappa)$. We will use the ordinal $(\eta^+)^{\text{Ult}(M,E)}$ as the index of E in sequences of extenders. The *trivial completion* of E is the $(\kappa, (\eta^+)^{\text{Ult}(M,E)})$ pre-extender G consisting of pairs (a, x) such that

$$a \in [(\eta^+)^{\text{Ult}(M,E)}]^{<\omega} \wedge x \subseteq [\kappa]^{\text{card}\, a} \wedge x \in M \wedge a \in i(x).$$

Then $E \restriction \alpha = G \restriction \alpha$, where $\alpha = \inf(\text{lh}\, E, \text{lh}\, G)$.

We now record some of the main properties of the extender sequences we shall consider:

DEFINITION 1.0.4. A sequence $\langle E_\beta \mid \beta \in S\rangle$ is *good at* α if it satisfies the following five clauses:

(1) $J_\alpha^{\vec{E}}$ is strongly acceptable,

and if $\alpha \in S$, then

(2) E_α is a (κ, α) pre-extender over $J_\alpha^{\vec{E}\restriction\alpha}$ for some κ such that $J_\alpha^{\vec{E}\restriction\alpha} \models \kappa^+$ exists,

(3) (bounded generators) E_α is the trivial completion of $E_\alpha \restriction \nu$, where ν is the natural length of E_α.

(4) (coherence) $i(\vec{E} \restriction \alpha) \restriction \alpha + 1 = \vec{E} \restriction \alpha$ where $i : J_\alpha^{\vec{E}\restriction\alpha} \to \mathrm{Ult}(J_\alpha^{\vec{E}\restriction\alpha}, E_\alpha)$ is the canonical embedding, and

(5) (closure under initial segment) Let ν be the natural length of E_α. If η is an ordinal such that $(\kappa^+)^{J_\alpha^{\vec{E}}} \leq \eta < \nu$ and η is the natural length of $E_\alpha \restriction \eta$, then one of (a) or (b) below holds:

(a) There is $\gamma < \alpha$ such that E_γ is the trivial completion of $E_\alpha \restriction \eta$.
(b) $\eta \in S$, and there is a $\gamma < \alpha$ such that $\pi(\vec{E} \restriction \eta)_\gamma$ is the trivial completion of $E_\alpha \restriction \eta$, where $\pi\colon \mathcal{J}_\eta^{\vec{E}\restriction\eta} \to \mathrm{Ult}(\mathcal{J}_\eta^{\vec{E}\restriction\eta}, E_\eta)$ is the canonical embedding.

DEFINITION 1.0.5. A potential premouse (ppm) is a structure of the form $\mathcal{J}_\beta^{\vec{E}}$, where $\vec{E}$ is good at all $\alpha \leq \beta$.

DEFINITION 1.0.6. A ppm $\mathcal{J}_\beta^{\vec{E}}$ is active if $\beta \in \mathrm{dom}\, \vec{E}$; otherwise it is passive.

DEFINITION 1.0.7. If $\mathcal{M} = \mathcal{J}_\beta^{\vec{E}}$ is active then $\nu^{\mathcal{M}}$ is the natural length of E_β.

Remarks. (a) Activity is determined by the similarity type of the ppm.

(b) Condition 3 implies every $\beta < \alpha$ is represented mod E_α by a function with support $\subseteq \nu^{\mathcal{M}}$, so that $E_\alpha \restriction \nu^{\mathcal{M}}$ determines all of E_α. We include the extra coordinates just so that the functions witnessing coherence will be trivial (essentially projections on a coordinate) which helps show coherence is preserved by Σ_0 ultrapowers. (Here $\mathcal{M} = \mathcal{J}_\alpha^{\vec{E}}$.)

(c) The ultrapowers in the definition are "Σ_0 ultrapowers", that is, formed using functions belonging to the model in question. Note $J_\alpha^{\vec{E}\restriction\alpha}$ is always passive, hence amenable, so that we can move its predicate.

Notice that since E_α is only a pre-extender, $\mathrm{Ult}(J_\alpha^{\vec{E}\restriction\alpha}, E_\alpha)$ may not be well-founded. However, $\alpha + 1 \subseteq \mathrm{wfp}\,(\mathrm{Ult})$, which is enough to make sense of conditions (3) and (4).

(d) Let $\vec{E}$ be good at α and $i : J_\alpha^{\vec{E}\restriction\alpha} \to \mathrm{Ult}(\mathcal{J}_\alpha^{\vec{E}\restriction\alpha}, E_\alpha)$ the canonical embedding. Let $\nu < \alpha$ be the natural length of E_α. By coherence, $J_\alpha^{i(\vec{E}\restriction\alpha)} = J_\alpha^{\vec{E}}$. Since $\alpha = \nu^+$ in $\mathrm{Ult}(J_\alpha^{\vec{E}\restriction\alpha}, E_\alpha)$, which is strongly acceptable, there are no cardinals

$> \nu$ in $J_\alpha^{i(\vec{E}\restriction\alpha)}$. So there are no cardinals $> \nu$ in $J_\alpha^{\vec{E}}$. The ordinal ν itself may be a successor ordinal. It is easy to see that if ν is a limit ordinal, then in fact ν is a cardinal, both in $J_\alpha^{\vec{E}}$ and $\mathrm{Ult}(J_\alpha^{\vec{E}\restriction\alpha}, E_\alpha)$.

(e) Let $\kappa = \mathrm{crit}\ E_\alpha$. By (3) there is a map of $(P(\kappa) \cap J_\nu^{\vec{E}}) \times [\nu]^{<\omega}$ onto α, the map being in $J_{\alpha+1}^{\vec{E}}$. Thus α is not a cardinal in $J_{\alpha+1}^{\vec{E}}$.

(f) For the good $\vec{E}$ we construct, E_α is an extender over $L[\vec{E} \restriction \alpha]$, which is strongly acceptable, and $\alpha = \nu^+$ in both $L[\vec{E} \restriction \alpha]$ and $\mathrm{Ult}(L[\vec{E} \restriction \alpha], E_\alpha)$. This in fact follows from the definition of goodness if we can iterate from $J_\alpha^{\vec{E}}$ via E_α and its images OR times (and preserve wellfoundedness).

(h) It might be hoped that alternative (b) of the initial segment condition could be dropped, but we suspect that if $L[\vec{E}]$ is to have a Woodin cardinal, or even lots of strong cardinals, one cannot demand this stronger version of the initial segment condition. The initial segment condition is crucial in the proof that the comparison process terminates (cf. §7). We need some form of it as an axiom on our extender sequences in order to get a decent theory going. Sy Friedman has suggested that it might be possible to eliminate this clause if the sequences are indexed by letting an extender be E_α where α is the double successor of the natural length of E in the ultrapower by E, instead of using the single successor as in this paper. We do not know whether this idea can be made to work.

Notice passive ppm are amenable. For active ppm we have a weaker property, which we call weak amenability.

DEFINITION 1.0.8. Let $J_\alpha^{\vec{E}}$ be an active ppm, and $\kappa = \mathrm{crit}\ E_\alpha$. We say $J_\alpha^{\vec{E}}$ is *weakly amenable* iff whenever $\langle A_\beta \mid \beta < \kappa\rangle \in J_\alpha^{\vec{E}}$ and $\forall\beta\exists n < \omega(A_\beta \subseteq [\kappa]^n)$ and $\eta < \alpha$, then

$$E_\alpha \cap ([\eta]^{<\omega} \times \{A_\beta \mid \beta < \kappa\}) \in J_\alpha^{\vec{E}}.$$

The proof of the next lemma is well-known (due to K. Kunen?).

Lemma 1.1. *Every active ppm is weakly amenable.*

PROOF. Let $\langle A_\beta \mid \beta < \kappa\rangle$ be as in the definition of weak amenability, and

$$i : J_\alpha^{\vec{E}\restriction\alpha} \to \mathrm{Ult}(J_\alpha^{\vec{E}\restriction\alpha}, E_\alpha) = \mathrm{Ult}$$

the canonical embedding. Let $F = E_\alpha \cap ([\eta]^{<\omega} \times \{A_\beta \mid \beta < \kappa\})$ where $\eta < \alpha$. Now

$$\langle i(A_\beta) \mid \beta < \kappa\rangle \in \mathrm{Ult}$$

so as $A_\beta = i(A_\beta) \cap [\kappa]^{<\omega}$,

$$F = \{(a, A_\beta) \mid a \in [\eta]^{<\omega} \wedge a \in i(A_\beta)\} \in \mathrm{Ult}.$$

By strong acceptability, since α is a cardinal in Ult, $F \in J_\alpha^{i(E\restriction\alpha)}$. By coherence, $F \in J_\alpha^{\vec{E}}$. □

Remark. Let ν be the natural length of E_α. It is easy to see that if $A \subset \nu$ and $A \in \text{Ult}(J_\alpha^{\vec{E}}, E_\alpha)$, then A can be computed from $E_\alpha \cap ([\nu]^{<\omega} \times \{A_\beta \mid \beta < \kappa\})$ for some sequence $\langle A_\beta \mid \beta < \kappa\rangle \in J_\nu^{\vec{E}}$ of subsets of $[\kappa]^n$, $n < \omega$. Thus in fact α is the least ordinal γ such that $E_\alpha \cap ([\nu]^{<\omega} \times \{A_\beta \mid \beta < \kappa\}) \in J_\gamma^{\vec{E}}$ for all such $\langle A_\beta \mid \beta < \kappa\rangle \in J_\nu^{\vec{E}}$. This is the motivation for condition (3) of good at α: we don't add an extender until we have weak amenability.

Remark. For $\xi < (\kappa^+)^{J_\alpha^{\vec{E}}}$ define γ_ξ to be the least ordinal γ such that $E_\alpha \cap \left([\nu]^{<\omega} \times J_\xi^{\vec{E}}\right) \in J_\gamma^{\vec{E}}$. The ordinals γ_ξ's are cofinal in α. To see this, let $A \in \text{Ult}(J_\alpha^{\vec{E}}, E_\alpha)$ be any subset of the natural length ν of E_α, and let $A = [a, f]_{E_\alpha}$. Then A can be computed from $(E_\alpha)_a \cap \{ A_\eta : \eta < \kappa \}$ where $A_\eta = \{ \bar{u} : \eta \in f(\bar{u}) \}$, so that if $f \in J_\xi^{\vec{E}}$ for $\xi < \kappa^+$ of $J_\alpha^{\vec{E}}$ then $A \in J_{\gamma_\xi+1}^{\vec{E}}$.

Since the sequence $(\gamma_\xi : \xi < (\kappa^+)^{J_\alpha^{\vec{E}}})$ is in $J_{\alpha+1}^{\vec{E}}$ it follows that $J_{\alpha+1}^{\vec{E}} \models \text{cf}(\alpha) = \text{cf}((\kappa^+)^{J_\alpha^{\vec{E}}})$.

§2. Fine Structure

This and the next section explain some basic facts about the fine structure of definability over potential premice. Most of the notions and results here are due to Dodd and Jensen ([DJ4]). There is one important difference, however: because E_α measures only sets in $J_\alpha^{\vec{E}\restriction\alpha}$, the hydras of [DJ4] disappear; moreover, we never have to iterate in order to extend measures. This simplifies the theory considerably.

Let $\mathcal{M} = \mathcal{J}_\beta^{\vec{E}}$. If $\nu^{\mathcal{M}}$ is a successor ordinal, we let $\gamma^{\mathcal{M}}$ be the witness to 5(a) or 5(b) with respect to the trivial completion of $E_\beta \restriction \nu^{\mathcal{M}} - 1$. That is, let G be the trivial completion of $E_\beta \restriction \eta$, where $\eta \leq \nu^{\mathcal{M}} - 1$ is the natural length of $E_\beta \restriction \nu^{\mathcal{M}}$.

Remark. If $\eta < \nu^{\mathcal{M}} - 1$, then $G = \dot{F}^{\mathcal{M}} \restriction \nu^{\mathcal{M}} - 1$. The proof uses the fact, which is not hard to prove, that if η is a limit of generators and is not itself a generator then η is not in S.

If 5 (a) of the definition of good at α holds, we set

$$\gamma^{\mathcal{M}} = \operatorname{lh} G = \text{the unique } \xi \in \operatorname{dom} \vec{E}^{\mathcal{M}} \text{ such that } G = \dot{E}_\xi^{\mathcal{M}}\,.$$

If 5(b) holds, then $\eta \in \operatorname{dom} \vec{E}^{\mathcal{M}}$ and G is on the sequence of $\operatorname{Ult}(\mathcal{M}, E_\eta^{\mathcal{M}})$. We set

$$\begin{aligned}(b, g) &= \text{the first pair } (a, f) \text{ in order of} \\ &\qquad \text{construction of } \mathcal{M} \text{ such that } G = [a, f]_{E_\eta}^{\mathcal{M}}, \text{ and} \\ \gamma^{\mathcal{M}} &= (\eta, b, g)\,.\end{aligned}$$

Definition 2.0.1. Let $\mathcal{M}$ be an active ppm, and $\kappa = \operatorname{crit} F$, where F is the new extender introduced by $\mathcal{M}$. Then

(a) $\mathcal{M}$ is type I iff $\nu^{\mathcal{M}} = (\kappa^+)^{\mathcal{M}}$

(b) $\mathcal{M}$ is type II iff $\nu^{\mathcal{M}}$ is a successor ordinal ($\nu^{\mathcal{M}} - 1 > (\kappa^+)^{\mathcal{M}}$ follows from the definition of $\nu^{\mathcal{M}}$),

(c) $\mathcal{M}$ is type III iff $\nu^{\mathcal{M}}$ is a limit ordinal and $\nu^{\mathcal{M}} > (\kappa^+)^{\mathcal{M}}$.

The definability hierarchy we study in this section is not appropriate for active type III ppm. The reason is explained at the beginning of §3, where we study a different hierarchy appropriate for such ppm. In this section we restrict our attention to ppm which are passive or active of types I or II.

Definition 2.0.2. $\mathcal{L}$ is the language of set theory with additional constant symbols $\dot{\nu}$, $\dot{\gamma}$, $\dot{\mu}$, additional 1-ary predicate symbol $\dot{E}$, and 3-ary predicate symbol $\dot{F}$.

If $\mathcal{M}$ is a ppm, then we interpret $\mathcal{L}$ in $\mathcal{M}$ as follows: Suppose that the structure $\mathcal{M} = (J_\alpha^{\vec{E}}, \in, \vec{E} \restriction \alpha, \tilde{E}_\alpha)$ is active. Then

$$\dot{E}^{\mathcal{M}} = \vec{E} \restriction \alpha\,, \quad \dot{F}^{\mathcal{M}} = \tilde{E}_\alpha\,,$$
$$\dot{\nu}^{\mathcal{M}} = \nu^{\mathcal{M}}\,, \qquad \dot{\mu}^{\mathcal{M}} = \text{crit } E_\alpha\,,$$
$$\dot{\gamma}^{\mathcal{M}} = \begin{cases} \gamma & \text{if } \mathcal{M} \text{ is type II} \\ 0 & \text{if } \mathcal{M} \text{ is type I} \end{cases}$$

Suppose $\mathcal{M} = (J_\alpha^{\vec{E}}, \in, \vec{E} \restriction \alpha)$ is passive. Then

$$\dot{E}^{\mathcal{M}} = \vec{E} \restriction \alpha\,, \ \dot{F}^{\mathcal{M}} = \emptyset$$
$$\dot{\mu}^{\mathcal{M}} = \dot{\nu}^{\mathcal{M}} = \dot{\gamma}^{\mathcal{M}} = 0\,.$$

An active ppm is not amenable with respect to its interpretation of $\dot{F}$. For this reason, the ultrapower of such a ppm via functions belonging to the ppm may not satisfy Łos' theorem for Σ_0 formulae of $\mathcal{L}$, and we must start our Levy-like hierarchy at a smaller class of formulae.

DEFINITION 2.0.3. $r\Sigma_0$ (or restricted Σ_0) is the smallest class of $\mathcal{L}$ formulae containing the atomic formulae of $\mathcal{L}$, all Σ_0 formulae of $\mathcal{L} - \{\dot{F}\}$, and closed under $\wedge$, $\vee, \neg$, and bounded quantification over finite sets. [That is, if $\theta(x, G, \bar{v})$ is $r\Sigma_0$, so are $\psi(G, \bar{v})$ and $\varphi(G, \bar{v})$, where $\psi(G, \bar{v}) =$ "G is finite $\wedge\ \exists x \in G\,\theta$" and $\varphi(G, \bar{v}) =$ "G is finite $\wedge\ \forall x \in G\,\theta$"]

DEFINITION 2.0.4. Let $\theta(\bar{v})$ be an $\mathcal{L}$ formula; then θ is $r\Sigma_1$ iff $\theta = \exists u \varphi(u, \bar{v})$, where φ is $r\Sigma_0$.

We could continue and define $r\Sigma_n$, $r\Pi_n$ for $n > 1$ by counting quantifier alternations. However, we shall reserve "$r\Sigma_n$" and "$r\Pi_n$" for $n > 1$, for different, more useful classes of formulae.

For $\mathcal{M}$ a ppm, $r\Sigma_n^{\mathcal{M}}$ is the class of relations on the universe of $\mathcal{M}$ definable over $\mathcal{M}$ by a $r\Sigma_n$ formula ($n = 0, 1$). If $\mathcal{M}$ is passive, then $r\Sigma_1^{\mathcal{M}} = \Sigma_1^{\mathcal{M}}$, the usual Levy class. The relativised classes $r\Sigma_n^{\mathcal{M}}(X)$, for $X \subseteq |\mathcal{M}|$, and $r\Sigma_n^{\mathcal{M}} = r\Sigma_n^{\mathcal{M}}(|\mathcal{M}|)$, are as usual.

Notice that $r\Sigma_1^{\mathcal{M}}$ is closed under $\exists$, $\wedge$, $\vee$, and bounded quantification over finite sets.

The following normal form theorem makes clearer what an $r\Sigma_1$ formulae can assert of an active ppm.

Lemma 2.1. *If $\varphi(\bar{v})$ is any $r\Sigma_1$ formula then there is a Σ_1 formula $\psi(a, b, \delta, \bar{v})$ of $\mathcal{L} - \{\dot{F}\}$ such that for all active ppm $\mathcal{M}$*

$$(*) \qquad \mathcal{M} \models \forall \bar{v}\,(\varphi(\bar{v}) \Leftrightarrow \exists a, b, \delta[\dot{F}(a, b, \delta) \wedge \psi(a, b, \delta, \bar{v})])\,.$$

Remark. That is, an $r\Sigma_1$ formula asserts there is a "small chunk" of the new extender having a Σ_1 in $(\mathcal{L} - \{\dot{F}\})$ property.

PROOF. We show first that if $\theta(\bar{v})$ is a $r\Sigma_0$ formula and $\varphi = \theta$ or $\varphi = \neg\theta$ then there is ψ as in (*). This is by induction on θ. E.g. for $\varphi = \neg\dot{F}(x, y, z)$, we have

$$\mathcal{M} \models \forall x, y, z\big(\neg\dot{F}(x,y,z) \Leftrightarrow \exists a, b, \delta\big[\dot{F}(a,b,\delta) \wedge \\ \wedge (z \text{ is not an ordinal } \vee y \text{ is not a function } \vee \\ \vee \operatorname{dom} y \neq \operatorname{dom} b \vee (y = b \wedge z = \delta \wedge a \neq x))\big]\big).$$

E. g. for the "$\forall x \in G$" step, let $\psi(a, b, \delta, x, G, \bar{v})$ be Σ_1 in $\mathcal{L}$ - $\{\dot{F}\}$; then we can rewrite

$$G \text{ finite } \wedge \forall x \in G \, \exists a, b, \delta[\dot{F}(a,b,\delta) \wedge \psi(a,b,\delta,x,G,\bar{v})]$$

as

$$(**) \quad \exists a, b, \delta\big[\dot{F}(a,b,\delta) \wedge \exists k < \omega \, \exists \text{ sequences } \bar{x}, \bar{a}, \bar{b}, \bar{\delta} \text{ of length } k \\ (G = \{x_1, \ldots, x_k\} \wedge \psi^*(\bar{x}, \bar{a}, \bar{\delta}, k, b))]$$

where ψ^* is the formula

$$\forall i \leq k\big(b_i \text{ is a function } \wedge \operatorname{dom} b_i = \operatorname{dom} b \wedge \operatorname{ran} b_i \subseteq \operatorname{ran} b \\ \wedge \delta_i < \delta \wedge a_i = a \cap ([\delta_i]^{<\omega} \times \operatorname{ran} b_i) \wedge \psi(a_i, b_i, \delta_i, x_i, G, \bar{v})\big).$$

Then the formula (**) is equivalent to

$$\exists a, b, \delta \exists k < \omega \, \exists \text{ sequences } \bar{x}, \bar{a}, \bar{b}, \bar{\delta} \text{ of length } k \\ \big[\dot{F}(a,b,\delta) \wedge G = \{x_1, \ldots, x_k\} \wedge \psi^*(\bar{x}, \bar{a}, \bar{\delta}, k, b)\big]$$

Thus every $r\Sigma_0$ formula φ can be put in the form given by (*). This easily implies the same is true of $r\Sigma_1$ φ. □

As a corollary, $r\Sigma_1$ satisfaction is $r\Sigma_1$, uniformly over all active ppm:

Corollary 2.2. *There is an $r\Sigma_1$ formula $\theta(v_0, v_1)$ such that whenever $\mathcal{M}$ is an active ppm*

$$\mathcal{M} \models \theta[i, \langle a_0 \cdots a_k\rangle] \quad \textit{iff } i \textit{ is the Gödel number of an } r\Sigma_1 \textit{ formula } \varphi \\ \textit{and } \mathcal{M} \models \varphi[a_0 \cdots a_k].$$

Of course, $r\Sigma_1 = \Sigma_1$ satisfaction over passive ppm is also uniformly $r\Sigma_1$.

Henceforth we will use the same letter for the Gödel number of a formula as we use for the formula, so that the displayed line in the corollary would start "$\mathcal{M} \models \theta[\varphi, \langle a_0 \cdots a_k\rangle]$."

Lemmas 2.1 and 2.2 hold also for type III active ppm, setting $\dot\gamma^{\mathcal{M}} = 0$ for $\mathcal{M}$ type III, but once again we have no use for this fact.

Remark. Let $\mathcal{M} = \mathcal{J}_\alpha^{\vec{E}}$ be active. We can construct an amenable structure $\mathcal{M}^*$ as follows. Define E_α^* to be the set of quadruples (γ, ξ, a, x) such that

$$(\alpha > \gamma > \nu^{\mathcal{M}}) \wedge (\dot\mu^{\mathcal{M}} < \xi < \dot\mu^{+\mathcal{M}}) \wedge \\ \wedge \left(E_\alpha \cap ([\nu^{\mathcal{M}}]^{<\omega} \times J_\xi^{\vec{E}}) \in J_\gamma^{\vec{E}}\right) \wedge \left((a, x) \in (E_\alpha \cap ([\gamma]^{<\omega} \times J_\xi^{\vec{E}}))\right).$$

Our remark on p. 10 shows that $\mathcal{M}^*$ is amenable; moreover one can easily see that $r\Sigma_1^{\mathcal{M}}$ is the usual Σ_1 over $\mathcal{M}^*$. This fact (which we didn't notice until we had written much of these notes) doesn't seem to simplify the definability analysis to follow. It might be used to give the analysis a more conventional look.

For any ppm $\mathcal{M}$, let $<^{\mathcal{M}}$ be the usual order of construction (so if $\mathcal{M} = \mathcal{J}_\alpha^{\vec{E}}$ and $\mathcal{N} = \mathcal{J}_\beta^{\vec{E}}$ where $\alpha < \beta$, then $<^{\mathcal{N}}$ end-extends $<^{\mathcal{M}}$).

Lemma 2.3. *There are Σ_1 formulae of $\mathcal{L} - \{\dot F\}$ defining uniformly over all ppm $\mathcal{J}_\alpha^{\vec{E}}$ the functions*

$$\beta \mapsto \mathcal{J}_\beta^{\vec{E}}, \quad \beta \mapsto <^{\mathcal{J}_\beta^{\vec{E}}}.$$

PROOF. See Dodd-Jensen [DJ1]. □

The formulae of the lemma are in $\mathcal{L} - \{\dot F\}$ hence $r\Sigma_1$. To apply [DJ1], notice every ppm is amenable with respect to its interpretation of $\dot E$ (and in fact $\dot E^{\mathcal{M}} \in \mathcal{M}$ if $\mathcal{M} = \mathcal{J}_{\alpha+1}^{\vec{E}}$).

Skolem terms and projecta.

Following Magidor and Silver, we shall introduce Skolem terms into our language as a convenience. The existence of the amenable structure $\mathcal{M}^*$, which we described following 2.2, shows that $r\Sigma_1^{\mathcal{M}}$ relations admit $r\Sigma_1$ uniformizations, and this means that we could avoid the Skolem terms if we cared to do so. There seems to be no great advantage to either approach. We shall also define the classes $r\Sigma_n^{\mathcal{M}}$ for $n > 1$, and introduce a predicate $\dot T_n$ related to $r\Sigma_n$ satisfaction. Like Magidor and Silver, we shall work directly with the classes $r\Sigma_n^{\mathcal{M}}$, rather than with Σ_1 definability over master code structures. However, $\dot T_n^{\mathcal{M}}$ is closely related to the nth master code of M, and its use in constructing $r\Sigma_{n+1}^{\mathcal{M}}$ relations makes this difference more apparent than real.

DEFINITION 2.3.1. $\mathcal{L}^+$ is $\mathcal{L}$ together with additional 2-ary predicate symbols $\dot T_n$ for $1 \le n < \omega$.

The interpretation $\dot T_n^{\mathcal{M}}$, for $\mathcal{M}$ a ppm, will be defined shortly.

DEFINITION 2.3.2. Let $\theta(\bar{v})$ be a formula of $\mathcal{L}^+$.

(a) θ is $r\Sigma_1$ iff θ is a formula of $\mathcal{L}$ which is $r\Sigma_1$ in our former sense.
(b) θ is $r\Sigma_{n+1}$ (where $n \geq 1$) iff there is a $r\Sigma_1$ formula $\psi(a, b, \bar{v})$ of $\mathcal{L}$ such that
$$\theta = \exists a \exists b(\dot{T}_n(a, b) \wedge \psi(a, b, \bar{v}))\,.$$

DEFINITION 2.3.3. Let $\varphi = \varphi(v_0, \ldots, v_k, v_{k+1})$ be a formula of $\mathcal{L}^+$. Then $\tau_\varphi(v_0, \ldots v_k)$ is the *basic Skolem term* associated to φ.

Given that we have interpreted φ in a ppm $\mathcal{M}$ (which we have not as yet done in general), we interpret τ_φ as follows:

$$\tau_\varphi^{\mathcal{M}}(a_0 \cdots a_k) = \begin{cases} <^{\mathcal{M}} \text{ least } b \text{ such that } \mathcal{M} \models \varphi[\bar{a}, b] & \text{if such } b \text{ exists} \\ 0 & \text{otherwise} \end{cases}$$

DEFINITION 2.3.4. For $n \geq 1$, Sk_n (the class of level n Skolem terms) is the smallest class which contains τ_φ for each $r\Sigma_n$ formula φ and is closed under composition.

DEFINITION 2.3.5. A formula φ of $\mathcal{L}^+$ is *generalized* $r\Sigma_n$ for $n \geq 1$ iff φ results from an $r\Sigma_n$ formula ψ by substituting terms in Sk_n for free variables in ψ.

(The substitution of τ into ψ must be such that no variable free in τ becomes bound in the resulting φ.)

We can now define the predicate $\dot{T}_n^{\mathcal{M}}$ for $\mathcal{M}$ a ppm; simultaneously, we define the nth projectum $\rho_n^{\mathcal{M}}$ of $\mathcal{M}$.

DEFINITION 2.3.6. Let $\mathcal{M}$ be a ppm and $n \geq 1$. Then

(a) $\mathrm{Th}_n^{\mathcal{M}}(X) = \{\langle \varphi, \bar{a} \rangle \mid \bar{a} \in X^{<\omega}$ and φ is a generalized $r\Sigma_n$ formula and $\mathcal{M} \models \varphi[\bar{a}]\}$,
(b) $\rho_n^{\mathcal{M}}$ is the least ordinal $\rho \subseteq |\mathcal{M}|$ such that $\mathrm{Th}_n^{\mathcal{M}}(\rho \cup \{q\}) \notin |\mathcal{M}|$ for some $q \in |\mathcal{M}|$,
(c) The predicate $\dot{T}_n^{\mathcal{M}}(a, b)$ holds if and only if $a = \langle \alpha, q \rangle$ for some $\alpha < \rho_n^{\mathcal{M}}$ and $q \in \mathcal{M}$, and $b = \mathrm{Th}_n^{\mathcal{M}}(\alpha \cup \{q\})$.

Of course, $\dot{T}_n$ is essentially 3-ary. It's present form is a relic of an earlier version of these notes, where we set $\dot{T}_n^{\mathcal{M}}(a, b) \Leftrightarrow b = \mathrm{Th}_n^{\mathcal{M}}(a)$. That earlier version led to a problem (showing $r\Sigma_n$ ultrapowers give rise to $r\Sigma_{n+1}$ elementary embeddings).

Remark. $\rho_n^{\mathcal{M}} = \mathrm{OR}^{\mathcal{M}}$ is possible.

The definition above is by induction on n, as (a) for n depends upon (c) for $1 \leq m < n$.

We define the classes $r\Sigma_n^{\mathcal{M}}$, $r\Pi_n^{\mathcal{M}}$, $r\Delta_n^{\mathcal{M}}$, for $\mathcal{M}$ a ppm, in the obvious way. The relativised and boldface versions of these classes are also defined in the obvious way.

Notice that if $\rho_n^{\mathcal{M}} \le \omega$, then in fact $\rho_n^{\mathcal{M}} = 0$, and $r\Sigma_{n+1}^{\mathcal{M}}$ trivializes. We are not interested in $r\Sigma_{n+1}^{\mathcal{M}}$ when $\rho_n^{\mathcal{M}} = 0$ (although we are definitely interested in $r\Sigma_n^{\mathcal{M}}$ in this case). We shall tacitly assume henceforth that, in any discussion of a class of the form $r\Sigma_{n+1}^{\mathcal{M}}$, $\rho_n^{\mathcal{M}} > 0$.

It is easy to see that $r\Sigma_n^{\mathcal{M}}$ is closed under $\exists$, $\wedge$, and $\vee$, and that $(r\Sigma_n^{\mathcal{M}} \cup r\Pi_n^{\mathcal{M}}) \subseteq r\Sigma_{n+1}^{\mathcal{M}}$, for any ppm $\mathcal{M}$. Moreover, the closure and inclusion are uniform over all ppm (there is a recursive translation procedure acting on formulae of the appropriate type). It is clear that $\neg \dot{T}_n^{\mathcal{M}}$ is $r\Sigma_{n+1}^{\mathcal{M}}$ in the parameter $\rho_n^{\mathcal{M}}$, uniformly over all $\mathcal{M}$.

It follows that the class of sets definable by $r\Sigma_{n+1}$ formulae would be unchanged if we modified definition 2.3.3 by allowing allowed any formula of the form $\exists a, b(\dot{T}_n(a,b) \wedge \psi(a,b,\bar{v}))$ where ψ is a boolean combination of $r\Sigma_n$ formulae. A similar argument shows that we could also have restricted ψ to be Σ_1 in $\mathcal{L} - \{\dot{F}, \dot{E}\}$.

Hulls.

DEFINITION 2.3.7. Let $\mathcal{M}$ be a ppm, $n \ge 1$, and $X \subseteq |\mathcal{M}|$. Then

$$S_n^{\mathcal{M}}(X) = \{\tau^{\mathcal{M}}(\bar{a}) \mid \tau \in \mathrm{Sk}_n \wedge \bar{a} \in X^{<\omega}\},$$
$$H_n^{\mathcal{M}}(X) = \pi'' S_n^{\mathcal{M}}(X) \quad \text{where } \pi \text{ is the transitive collapse,}$$
$$\mathcal{H}_n^{\mathcal{M}}(X) = (H_n^{\mathcal{M}}(X), \in, \pi''(\dot{E}^{\mathcal{M}}), \pi''(\dot{F}^{\mathcal{M}})).$$

(The last predicate occurs only if $\mathcal{M}$ is active).

We shall show $\mathcal{H}_n^{\mathcal{M}}(X)$ is a ppm. To this end, recall the Q formulae of [DJ4]. One virtue of these formulae is that they go down under Σ_1 embeddings and up under cofinal Σ_0 embeddings. We now define the appropriate analog in our situation.

DEFINITION 2.3.8. Let $\mathcal{M}$ be a ppm, and $\pi : \mathcal{M} \to \mathcal{P}$ be an $r\Sigma_0$ embedding, with $\mathcal{P}$ transitive. We say π is *cofinal* iff

(a) $\forall\, y \in |\mathcal{P}|\, \exists x\, (y \subseteq \pi(x))$, and

(b) $\pi''(\dot{\mu}^+)^{\mathcal{M}}$ is cofinal in $\pi((\dot{\mu}^+)^{\mathcal{M}})$.

Recall here $\dot{\mu}^{\mathcal{M}} = 0$ if $\mathcal{M}$ is passive, so that (b) is trivially true then. If $\mathcal{M}$ is active, $\dot{\mu}^{\mathcal{M}} = \mathrm{crit}\ \dot{F}^{\mathcal{M}}$.

DEFINITION 2.3.9. An rQ formula is one of the form:

$$\forall x \forall \theta < \dot{\mu}^+\, \exists y \exists \nu\, (x \subseteq y \wedge (\theta \le \nu < \dot{\mu}^+) \wedge \varphi(y, \nu, \bar{u}))$$

where φ is $r\Sigma_1$ and does not have x or θ free.

Interpreted in a ppm $\mathcal{M}$, an rQ formula asserts that, in the product order on $(\dot{\mu}^+)^{\mathcal{M}} \times |\mathcal{M}|$ determined by the inclusion order on the factors, there are cofinally many pairs (ν, y) with an $r\Sigma_1^{\mathcal{M}}$ property. (If $\mathcal{M}$ is passive, this reduces to asserting that, under inclusion, there are cofinally many $y \in |\mathcal{M}|$ with an $r\Sigma_1^{\mathcal{M}}$ property.) So we have clearly

Lemma 2.4. *Let $\varphi(\bar{v})$ be an rQ formula, and $\pi : \mathcal{M} \to \mathcal{P}$. Then*
(a) *If π is an $r\Sigma_1$ embedding and $\mathcal{P} \models \varphi[\pi(\bar{a})]$, then $\mathcal{M} \models \varphi[\bar{a}]$.*
(b) *If π is a cofinal $r\Sigma_0$ embedding and $\mathcal{M} \models \varphi[\bar{a}]$, then $\mathcal{P} \models \varphi[\pi(\bar{a})]$.*

The preservation properties given in 2.4 are interesting because one can say with an rQ sentence: "I am a (passive/active) ppm".

Lemma 2.5. *There are rQ sentences φ_1, φ_2, φ_3 such that if $\mathcal{M}$ is a transitive $\mathcal{L}$-structure, then*
(a) *$\mathcal{M} \models \varphi_1$ holds if and only if $\mathcal{M}$ is passive and $\langle \dot{\mu}^{\mathcal{M}}, \dot{\nu}^{\mathcal{M}}, \dot{\gamma}^{\mathcal{M}} \rangle = \langle \mu^{\mathcal{M}}, \nu^{\mathcal{M}}, \gamma^{\mathcal{M}} \rangle$;*
(b) *$\mathcal{M} \models \varphi_2$ holds if and only if $\mathcal{M}$ is active type* I *and $\langle \dot{\mu}^{\mathcal{M}}, \dot{\nu}^{\mathcal{M}}, \dot{\gamma}^{\mathcal{M}} \rangle = \langle \mu^{\mathcal{M}}, \nu^{\mathcal{M}}, \gamma^{\mathcal{M}} \rangle$;*
(c) *$\mathcal{M} \models \varphi_3$ holds if and only if $\mathcal{M}$ is active type* II *and $\langle \dot{\mu}^{\mathcal{M}}, \dot{\nu}^{\mathcal{M}}, \dot{\gamma}^{\mathcal{M}} \rangle = \langle \mu^{\mathcal{M}}, \nu^{\mathcal{M}}, \gamma^{\mathcal{M}} \rangle$.*

PROOF. We construct φ_3, the other constructions being slightly simpler.

We shall use the fact that every $r\Pi_1$ formula can be put in rQ form; we leave it to the reader to check this.

By Dodd-Jensen [DJ4] there is an rQ sentence θ_1 of $\mathcal{L} - \{\dot{F}\}$ whose transitive models $\mathcal{M}$ are precisely those of the form $(J_\alpha^{\dot{E}^{\mathcal{M}}}, \in, \dot{E}^{\mathcal{M}}, \dot{F}^{\mathcal{M}})$. We add to θ_1 the $r\Pi_1$ sentence of $\mathcal{L} - \{\dot{F}\}$ stating that $\mathcal{M}$ is strongly acceptable and that $P(\dot{\mu})^{\mathcal{M}} \subseteq (J_{\dot{\nu}})^{\mathcal{M}}$.

Now we define a rQ sentence θ_2 asserting that $\dot{F}^{\mathcal{M}}$ codes a pre-extender over $\mathcal{M}$. The pre-extender coded is $\bigcup\{a \mid \exists b, \delta\, \dot{F}^{\mathcal{M}}(a, b, \delta)\}$. The formula θ_2 is the conjunction of the formulas (i–vi) below:
(i) "there are cofinally many (θ, γ) in the product order on $\dot{\mu}^+ \times \mathrm{OR}$ such that $\exists a\, \exists b(\dot{F}(a, b, \gamma) \wedge b$ is a function from $\dot{\mu}$ onto $P(\dot{\mu}) \cap J_\theta^{\dot{E}})$."
(ii) $\forall a, b, \gamma, a', b', \gamma'$ (if $\dot{F}(a, b, \gamma) \wedge \gamma' \leq \gamma \wedge b'$ is a function with dom $b' = \dot{\mu}$ and ran $b' \subseteq$ ran $b \wedge a' = a \cap ([\gamma']^{<\omega} \times \text{ran } b')$, then $\dot{F}(a', b', \gamma')$).
(iii) $\dot{F}(a, b, \gamma) \Rightarrow \gamma \in \mathrm{OR}$ and $b : \dot{\mu} \to P(\dot{\mu})$ and $a \subseteq [\gamma]^{<\omega} \times \text{ran } b$ and, letting $E_c = \{x \mid (c, x) \in a\}$, the E_c's are compatible $\dot{\mu}$-complete measures on $[\dot{\mu}]^{\text{card } c}$ "as far as sets in ran b are concerned".
(iv) $\dot{F}(a, b, \gamma) \wedge \dot{F}(a', b, \gamma) \Rightarrow a = a'$.
(v) (Normality) $(\forall f : [\dot{\mu}]^n \to \dot{\mu})(\forall b : \dot{\mu} \to (P([\dot{\mu}]^n) \cup P([\dot{\mu}]^{n+1})))$ [if b is f-closed then $\forall a, \delta(\dot{F}(a, b, \delta) \Rightarrow a = \langle E_c \mid c \in [\delta]^{<\omega} \rangle$ is normal with

respect to f)] (where "b is f-closed" stands for the formula: $\forall A \in \operatorname{ran} b \cap P([\dot{\mu}]^n)\forall i < n\{\langle\alpha_1 \ldots \alpha_i, \beta, \alpha_{i+1} \ldots \alpha_n\rangle \mid \langle\alpha_1 \ldots \alpha_n\rangle \in A \wedge \beta = f(\alpha_1 \ldots \alpha_n)\} \in \operatorname{ran} b$).

So far, condition (i) is rQ while (ii)-(v) are actually $r\Pi_1$, and we have asserted enough to ensure that $\mathrm{Ult}(\mathcal{M}, F)$ makes sense whenever $\mathcal{M} \models$ (i)-(v), where $F = \bigcup\{a \mid \dot{F}^{\mathcal{M}}(a, b, \delta)$ for some $b, \delta\}$. Normality guarantees $\mathrm{OR}^{\mathcal{M}} \subseteq \omega fp(\mathrm{Ult})$, but we must have $\mathrm{OR}^{\mathcal{M}} \in \mathrm{wfp}(\mathrm{Ult})$ for pre-extenderhood. From condition 3 of goodness at α ($\alpha = \mathrm{OR}^{\mathcal{M}}$), we know that we want to assert that $[\{\dot{\nu}\}^{\mathcal{M}}, f]_F^{\mathcal{M}} = \alpha$ where $f(\beta) = (\beta^+)^{\mathcal{M}}$ for $\beta < \dot{\mu}^{\mathcal{M}}$. The next clauses in θ_2 do this.

(vi) $\forall$ ordinals $\delta > \dot{\nu}$ $\forall\gamma > \delta$ $\forall a, b$ (if $\dot{F}(a, b, \gamma)$ and $\{(\alpha, \beta) \mid J_{\dot{\mu}}^{\dot{E}} \models \operatorname{card} \beta \leq \alpha\} = x$ is in $\operatorname{ran} b$, then $(\{\dot{\nu}, \delta\}, x) \in a$).

We have to say finally that there is no function "between" $f(\beta) = \beta^+$ on the $\dot{\nu}$th coordinate and the projection functions on arbitrary coordinates.

(vii) For cofinally many pairs (θ, γ) in the product order on $\dot{\mu}^+ \times \mathrm{OR}$ there are a, b and δ such that

$$\dot{F}(a, b, \delta) \wedge \delta > \gamma \wedge \forall n < \omega \left(P([\dot{\mu}]^n) \cap J_\theta^{\dot{E}} \subseteq \operatorname{ran} b\right)$$

and for all functions $f \in J_\theta^{\dot{E}}$ such that $f : [\dot{\mu}]^n \to \dot{\mu}$, and for all $c \in [\gamma]^{<\omega}$ such that $c = \{\eta_1 \cdots \eta_n\}$ for some ordinals $\eta_1 < \cdots < \eta_n$ with $\eta_i = \dot{\nu}$, and

$$(c, \{(\alpha_1 \ldots \alpha_n) \mid f(\alpha_1 \ldots \alpha_n) < (\alpha_i^+)^{J_{\dot{\mu}}^{\dot{E}}}\}) \in a$$

there is an ordinal ξ such that $\gamma \leq \xi < \delta$ and

$$(c \cup \{\xi\}, \{(\alpha_1 \cdots \alpha_{n+1}) \mid f(\alpha_1 \cdots \alpha_n) < \alpha_{n+1}\}) \in a\,.$$

The formula in (vii) is rQ. To see that if $\mathcal{M}$ satisfies (i–vii) then $f(\beta) = (\beta^+)^{\mathcal{M}}$, on the $\dot{\nu}^{\mathcal{M}}$ coordinate, represents $\mathrm{OR}^{\mathcal{M}}$ in Ult, notice that as $\dot{\mu}^{\mathcal{M}}$ is a cardinal of $\mathcal{M}$, strong acceptability implies $((\alpha_i^+)^{J_{\dot{\mu}}^{\dot{E}}})^{\mathcal{M}} = (\alpha_i^+)^{\mathcal{M}}$ for $\alpha_i < \dot{\mu}^{\mathcal{M}}$. We leave to the reader the not entirely trivial fact that any active ppm satisfies (vii).

Let θ_2 be the conjunction of (i)-(vii). If $\mathcal{M}$ satisfies $\theta_1 \wedge \theta_2$, then $\mathcal{M}$ satisfies conditions 1, 2, and part of 3 of good at α, for $\alpha = \mathrm{OR}^{\mathcal{M}}$. We capture the rest of condition 3 with θ_3:

θ_3: There are cofinally many $\gamma \in \mathrm{OR}$ such that $\exists a, b, \delta(\dot{F}(a, b, \delta) \wedge \delta > \gamma \wedge \exists f : [\dot{\mu}]^n \to \dot{\mu}\, \exists c \in [\dot{\nu}]^n$ such that $\dot{\nu} - 1 \in c$ and

$$(c \cup \{\gamma\}, \{(\alpha_1 \ldots \alpha_n, \beta) \mid f(\alpha_1 \ldots \alpha_n) = \beta \wedge J_{\dot{\mu}}^{\dot{E}} \models \operatorname{card}(\beta) \leq \alpha_n\}) \in a\,.$$

Moreover, $\dot{\nu} - 1$ is a generator of $\dot{F}$; that is $\forall a, b, \delta\ \forall f : [\dot{\mu}]^n \to \dot{\mu}$

$$\forall c \subseteq \dot{\nu} - 1\, (c \cup \{\dot{\nu} - 1\},\ \{(\alpha_1 \ldots \alpha_n, \beta) \mid f(\alpha_1 \ldots \alpha_n) = \beta\}) \notin a\,.$$

The formula θ_3 is the conjunction of an rQ sentence and an $r\Pi_1$ sentence, so θ_3 in rQ. One can check that if $\mathcal{M} \models \theta_3$, $\dot{\nu}^{\mathcal{M}} - 1$ is the largest generator of $\dot{F}^{\mathcal{M}}$. Notice here that if $\gamma \geq \dot{\nu}$ satisfies the displayed clause of θ_3, then there are no generators between γ and $\dot{\nu}$.

Recall that we are working with a type II ppm $\mathcal{M}$, so that $\dot{\nu}^{\mathcal{M}} - 1$ exists.

We can capture coherence, which is condition 4 of good at α, with an $r\Pi_1$ sentence θ_4: θ_4 just says $\forall a, b, \delta\ (\dot{F}(a, b, \delta) \Rightarrow$ "a is coherent as far as sets in b go"). We omit further detail.

Condition 5 is a disjunction of two possibilities, (a) and (b), and we accordingly set $\theta_5 = \psi_1 \vee \psi_2$. The formula ψ_1, asserting that clause 5a holds, is "$\dot{\gamma} \geq \dot{\nu} - 1$ and $\dot{\gamma} \in \operatorname{dom} \dot{E}$ and $\forall a, b(\dot{F}(a, b, \dot{\nu} - 1) \Rightarrow a \subseteq \dot{E}_{\dot{\gamma}})$ and $\forall \xi < \dot{\gamma}$ (ξ a generator of $\dot{E}_{\dot{\gamma}} \Rightarrow \xi < \dot{\nu} - 1$)." The formula ψ_1 is $r\Pi_1$ (its third conjunct is the only one not Σ_0 in $\mathcal{L} - \{\dot{F}\}$).

The formula ψ_2, asserting that clause 5b holds, says that $\dot{\gamma} = (\eta, b, g)$, where if we set $G = [b, g]_{\dot{E}_\eta}$ then η is the natural length of G and is in $\operatorname{dom}(\dot{E})$, the conjunction of the following three formulas holds:

$$\forall a, b(\dot{F}(a, b, \dot{\nu} - 1) \Rightarrow a \subseteq G)$$

$$g(\bar{u}) \text{ is on } \dot{E} \text{ for } (\dot{E}_\eta)_b \text{ a.e. } \bar{u}$$

$$\forall \xi < \operatorname{lh} G(\xi \text{ a generator of } G \Rightarrow \xi < \dot{\nu} - 1)$$

and finally $G \neq [a, f]_{\dot{E}_\eta}$ whenever (a, f) is constructed before (b, g). We leave it to the reader to see that the formula ψ_2 is also $r\Pi_1$.

The formula $\theta_5 = \psi_1 \vee \psi_2$ captures (5) for the "last" proper initial segment of $\dot{F}^{\mathcal{M}}$. Together with the Π_0 in $\mathcal{L} - \{\dot{F}\}$ assertion that $\dot{E}^{\mathcal{M}}$ is good at all $\beta < \alpha$, θ_5 captures (5).

Let φ be the Π_1 assertion that $\dot{E}^{\mathcal{M}}$ is good at all $\beta < \mathrm{OR}^{\mathcal{M}}$. Then $\varphi \wedge \bigwedge_{i \leq 5} \theta_i$ is the desired rQ sentence. This completes the proof of 2.5. □

Corollary 2.6. *Let $\mathcal{M}$ be a ppm which is passive or active of types* I *or* II. *Then*

(a) *if $\pi : \mathcal{H} \to \mathcal{M}$ is an $r\Sigma_1$ embedding, then $\mathcal{H}$ is a ppm of the same type as $\mathcal{M}$ and $\pi(\mu^{\mathcal{H}}) = \mu^{\mathcal{M}}$, $\pi(\nu^{\mathcal{H}}) = \nu^{\mathcal{M}}$, and $\pi(\gamma^{\mathcal{H}}) = \gamma^{\mathcal{M}}$,*

(b) *if $\pi : \mathcal{M} \to \mathcal{P}$ is either a cofinal $r\Sigma_0$ embedding or a "Σ_1 over $r\Sigma_1$" embedding (see the proof of* 2.7 *for the definition) then $\mathcal{P}$ is a ppm of the same type as $\mathcal{M}$, and $\pi(\mu^{\mathcal{M}}) = \mu^{\mathcal{P}}$, $\pi(\nu^{\mathcal{M}}) = \nu^{\mathcal{P}}$, and $\pi(\gamma^{\mathcal{M}}) = \gamma^{\mathcal{P}}$.*

The natural embedding $\pi : \mathcal{H}_n^{\mathcal{M}}(X) \to \mathcal{M}$ is clearly $r\Sigma_1$, so it follows that $\mathcal{H}_n^{\mathcal{M}}(X)$ is a ppm of the same type as $\mathcal{M}$. The next lemma shows that in certain circumstances π in fact preserves generalized $r\Sigma_n$ formulae.

Lemma 2.7. *Let $\mathcal{M}$ be a ppm which is passive or active type I or II. Let $\mathcal{H} = \mathcal{H}_n^{\mathcal{M}}(X)$, where $X \subseteq |\mathcal{M}|$ and $n \geq 1$. Suppose that if $n \geq 2$, then*

$$\rho_{n-1}^{\mathcal{M}} < \mathrm{OR}^{\mathcal{M}} \Rightarrow \exists q \in X\,(\mathrm{Th}_{n-1}^{\mathcal{M}}(\rho_{n-1}^{\mathcal{M}} \cup \{q\}) \notin |\mathcal{M}|)$$

and if $n \geq 3$, then

$$\rho_{n-2}^{\mathcal{M}} < OR^{\mathcal{M}} \Rightarrow \rho_{n-2}^{\mathcal{M}} \in X \wedge \exists q \in X\,(\mathrm{Th}_{n-2}^{\mathcal{M}}(\rho_{n-2}^{\mathcal{M}} \cup \{q\}) \notin |\mathcal{M}|).$$

Let $\pi : \mathcal{H} \to \mathcal{M}$ be the inverse of the collapse. Then

(a) *$\mathcal{H} \models \varphi[\bar{a}]$ iff $\mathcal{M} \models \varphi[\pi(\bar{a})]$ for φ generalized $r\Sigma_n$ and $\bar{a} \in H$*
(b) *for $1 \leq i \leq n-2$*

$$\rho_i^{\mathcal{M}} = \begin{cases} \mathrm{OR}^{\mathcal{M}} & \text{if } \rho_i^{\mathcal{H}} = OR^{\mathcal{H}} \\ \pi(\rho_i^{\mathcal{H}}) < OR^{\mathcal{M}} & \text{if } \rho_i^{\mathcal{H}} < \mathrm{OR}^{\mathcal{H}} \end{cases}$$

(c) $$\rho_{n-1}^{\mathcal{H}} = \begin{cases} \text{the least } \alpha \text{ such that } \pi(\alpha) \geq \rho_{n-1}^{\mathcal{M}} \\ OR^{\mathcal{H}} \text{ if no such } \alpha \text{ exists.} \end{cases}$$

PROOF. For $i \geq 0$ and $k \geq 1$, we say a formula φ is Σ_k over (generalized) $r\Sigma_i$ iff

$$\varphi = \exists v_1 \forall v_2 \exists v_3 \cdots Q_k v_k \psi$$

where ψ is a Boolean combination of (generalized) $r\Sigma_i$ formulae and $Q_k = \exists$ or $\forall$ as appropriate. (Here generalized $r\Sigma_0 = r\Sigma_0$.)

We show by induction on i that for $0 \leq i \leq n-1$

(i) If φ is Σ_{n-i} over generalized $r\Sigma_i$, then for all $\bar{a} \in H$

$$\mathcal{H} \models \varphi[\bar{a}] \Leftrightarrow \mathcal{M} \models \varphi[\pi(\bar{a})],$$

(ii) for $1 \leq i \leq n-1$ and $a, b \in H$

$$\mathrm{Th}_i^{\mathcal{H}}(a) = b \text{ iff } \mathrm{Th}_i^{\mathcal{M}}(\pi(a)) = \pi(b),$$

(iii) for $1 \leq i \leq n-2$, (b) of the statement of the lemma holds, and for $i = n-1$, (c) holds

(iv) if φ is generalized $r\Sigma_{i+1}$, then for all $\bar{a} \in H$

$$\mathcal{H} \models \varphi[\bar{a}] \Leftrightarrow \mathcal{M} \models \varphi[\pi(\bar{a})].$$

Proof of (i). If θ is Σ_{n-i} over generalized $r\Sigma_i$, then the translation procedure mentioned earlier gives us an $r\Sigma_n$ formula θ^* which is equivalent to θ over all ppm. As $\operatorname{ran}\pi$ is closed under $\tau_{\theta^*}^{\mathcal{M}}$, we see that if $\exists x(\mathcal{M} \models \theta[x, \pi(b)])$, then $\exists x \in \operatorname{ran} \pi(\mathcal{M} \models \theta[x, \pi(b)])$. But now π is elementary with respect to all generalized $r\Sigma_i$ formulae (trivially if $i = 0$, or by the induction hypothesis (iv) if $i > 0$). So the usual induction on the length of the quantifier prefix in φ gives (i).

Proof of (ii). First we observe that for any $i \geq 1$ there is a Π_1 over $r\Sigma_i$ formula $\sigma(v_0, v_1)$ such that for any ppm $\mathcal{P}$

$$\mathcal{P} \models \sigma[a, b] \Leftrightarrow \mathrm{Th}_i^{\mathcal{P}}(a) = b\,.$$

To see this, notice first that there is a recursive function associating to each term $\tau \in \mathrm{Sk}_i$ a Σ_1 over $r\Sigma_i$ formula θ_τ such that $\tau^{\mathcal{P}}[\bar{a}] = b$ iff $\mathcal{P} \models \theta_\tau[\bar{a}, b]$, for all ppm $\mathcal{P}$. For basic τ, say $\tau = \tau_\varphi$, let $\theta_\tau(\bar{u}, v)$ be the formula

$$\big(\varphi(\bar{u}, v) \wedge \forall w < v \, \neg\varphi(\bar{u}, w)\big) \vee \big(v = 0 \wedge \forall w \, \neg\varphi(\bar{u}, w)\big)\,.$$

In this case θ_τ is a Boolean combination of $r\Sigma_i$ formulae. The extension of $\tau \mapsto \theta\tau$ to all of Sk_i is obvious. Notice second that $r\Sigma_i$ satisfaction is uniformly $r\Sigma_i$ over all ppm.

It then follows that generalized $r\Sigma_i$ satisfaction is uniformly Σ_1 over $r\Sigma_i$, as well as uniformly Π_1 over $r\Sigma_i$, over all ppm. This gives us the desired formula σ.

Clause (ii) follows easily from (i) and the existence of σ.

proof of (iii). We first prove clause (b) for $i \leq n - 3$. Consider for example the first equivalence. The statement "$\rho_i^{\mathcal{M}} = \mathrm{OR}^{\mathcal{M}}$" can be expressed

$$\mathcal{M} \models \forall \alpha \in \mathrm{OR} \, \forall q \exists b \, \sigma(\alpha \cup \{q\}, b)$$

where σ is the formula asserting that $b = \mathrm{Th}_i^{\mathcal{P}}(a)$ from part (ii). This sentence is Π_3 over $r\Sigma_i$, so true in $\mathcal{M}$ iff true in $\mathcal{H}$ as $i \leq n - 3$ and we have induction hypothesis (i) at i. A similar calculation gives the second equivalence.

Clause (b) for $i = n - 2$ comes from a similar calculation. If $\rho_i^{\mathcal{M}} = \mathrm{OR}^{\mathcal{M}}$ then as we have just seen this is expressible by a Π_3 over $r\Sigma_i$ sentence which, since true in $\mathcal{M}$, must go down to $\mathcal{H}$ by induction hypothesis (i). If $\rho_i^{\mathcal{M}} < \mathrm{OR}^{\mathcal{M}}$, then by hypothesis $\rho_i^{\mathcal{M}}$ and a suitable parameter p are in $\mathrm{ran}(\pi)$. We get $\mathcal{M} \models \forall b \neg\sigma(\rho_i^{\mathcal{M}} \cup \{p\}, b)$, which is Π_2 over $r\Sigma_i$ and thus goes down to $\mathcal{H}$, showing $\rho_i^{\mathcal{H}} \leq \pi^{-1}(\rho_i^{\mathcal{M}})$. The second implication comes from a similar calculation.

Finally, in the case $i = n - 1$ we must prove (c).

Let $\pi(\alpha) < \rho_{n-1}^{\mathcal{M}}$; we claim $\alpha < \rho_{n-1}^{\mathcal{H}}$. For let $q \in |\mathcal{H}|$. Then

$$\begin{aligned}\mathrm{Th}_{n-1}^{\mathcal{M}}(\pi(\alpha) \cup \{\pi(q)\}) = \text{unique } c \text{ such that } \exists a, b(\dot{T}_{n-1}^{\mathcal{M}}(a, b) \wedge \\ a = \pi(\alpha) \cup \{\pi(q)\} \wedge b = c)\end{aligned}$$

so we can find $b \in |\mathcal{H}|$ such that

$$\mathrm{Th}_{n-1}^{\mathcal{M}}(\pi(\alpha) \cup \{\pi(q)\}) = \pi(b)\,.$$

But then $\mathrm{Th}^{\mathcal{H}}_{n-1}(\alpha \cup \{q\}) = b$ by (ii), so $\mathrm{Th}^{\mathcal{H}}_{n-1}(\alpha \cup \{q\}) \in |\mathcal{H}|$, and as q was arbitrary, $\alpha < \rho^{\mathcal{H}}_{i-1}$.

On the other hand, if $\pi(\alpha) \geq \rho^{\mathcal{M}}_{i-1}$, then by hypothesis we have a $p \in \operatorname{ran} \pi$ such that

$$\mathrm{Th}^{\mathcal{M}}_{n-1}(\rho^{\mathcal{M}}_{n-1} \cup \{p\}) \notin |\mathcal{M}|\,.$$

Let $\pi(q) = p$. Then $\mathrm{Th}^{\mathcal{H}}_{n-1}(\alpha \cup \{q\}) \notin |\mathcal{H}|$, so $\alpha \geq \rho^{\mathcal{H}}_{n-1}$.

Finally, we prove (iv) at i. Notice first that

$$\dot{T}^{\mathcal{H}}_i(a,b) \quad \text{iff} \quad \dot{T}^{\mathcal{M}}_i(\pi(a),\pi(b))\,.$$

For suppose $\dot{T}^{\mathcal{H}}_i(a,b)$. Then $a = \langle \alpha, q\rangle$ where $\alpha < \rho^{\mathcal{H}}_i$, and $b = \mathrm{Th}^{\mathcal{H}}_i(\alpha \cup \{q\})$. By (ii), $\pi(b) = \mathrm{Th}^{\mathcal{M}}_i(\pi(\alpha) \cup \{\pi(q)\})$, and by (iii) $\pi(\alpha) < \rho^{\mathcal{M}}_i$. Thus $\dot{T}^{\mathcal{M}}_i(\pi(a),\pi(b))$. The converse is equally easy.

It follows at once that π is $r\Sigma_{n+1}$ elementary. Suppose for example that $\mathcal{M} \models \exists a, b(\dot{T}_i(a,b) \wedge \psi(a,b,\pi(\bar{c})))$. Then applying the proper term in Sk_n to $\pi(\bar{c})$ we get $a, b \in$ ran π such that $\dot{T}^{\mathcal{M}}_i(a,b) \wedge \psi(a,b,\pi(\bar{c}))$, and we're done.

As the graph of any basic $\tau \in \mathrm{Sk}_{i+1}$ is definable by a boolean combination of $r\Sigma_{i+1}$ formulae, uniformly over all ppm, we see now that π is generalized $r\Sigma_{i+1}$ elementary.

This completes the proof of lemma 2.7.

Standard parameters and cores.

DEFINITION 2.7.1. A *parameter* of $\mathcal{M}$ is a sequence $\langle \alpha_0, \ldots, \alpha_k\rangle$ of ordinals of $\mathcal{M}$ such that $\alpha_0 > \alpha_1 > \cdots > \alpha_k$.

DEFINITION 2.7.2. $<_{\mathrm{lex}}$ is the lexicographic wellordering of all parameters (i.e. of all descending sequences of ordinals).

DEFINITION 2.7.3. Given a ppm $\mathcal{M}$ with $\rho^{\mathcal{M}}_k < \mathrm{OR}^{\mathcal{M}}$ and given $q \in |\mathcal{M}|$, the *kth standard parameter of* $(\mathcal{M}, q)$ is the $<_{\mathrm{lex}}$ least parameter p of $\mathcal{M}$ such that $\mathrm{Th}^{\mathcal{M}}_k(\rho^{\mathcal{M}}_k \cup \{p,q\}) \notin |\mathcal{M}|$.

We now define two useful properties a parameter might possess, solidity and universality. We shall eventually show that the appropriate standard parameters associated to the levels of the model we shall construct are solid and universal.

Solid parameters.

DEFINITION 2.7.4. Let $r = \langle \alpha_0 \cdots \alpha_\ell\rangle$ be a descending sequence of ordinals, $\mathcal{M}$ a ppm which is passive or active of types I or II, and $q \in |\mathcal{M}|$, and $1 \leq k < \omega$. We say r is k *solid over* $(\mathcal{M}, q)$ iff for all $i \leq \ell$

$$\mathrm{Th}^{\mathcal{M}}_k(\alpha_i \cup \{\langle \alpha_0 \cdots \alpha_{i-1}\rangle, q\}) \in |\mathcal{M}|\,.$$

We are interested in the case that r is the kth standard parameter of $(\mathcal{M}, q)$. Notice that in this case $\alpha_\ell \geq \rho_k^{\mathcal{M}}$, and for any finite $s \subseteq \alpha_i$, $\mathrm{Th}_k^{\mathcal{M}}(\rho_k^{\mathcal{M}} \cup s \cup \{\langle \alpha_0 \cdots \alpha_{i-1} \rangle, q\}) \in |\mathcal{M}|$ simply by the $<_{\mathrm{lex}}$ minimality of r. Solidity is the uniform version of this closure property of $\mathcal{M}$: $\langle \alpha_0 \ldots \alpha_\ell \rangle$ is k-solid over $(\mathcal{M}, q)$ iff $\pi \in \mathcal{M}$, where $\pi(s,i) = \mathrm{Th}_k^{\mathcal{M}}(\rho_k^{\mathcal{M}} \cup s \cup \{\langle \alpha_0 \ldots \alpha_{i-1} \rangle, q\})$, for all $i \leq \ell$ and finite $s \subseteq \alpha_i$.

Solidity is useful because it is easier to show solidity is preserved by the appropriate embeddings than to show standardness is.

Universal parameters.

DEFINITION 2.7.5. Let $\mathcal{M}$ be a ppm, $q \in |\mathcal{M}|$, r a parameter of $\mathcal{M}$, and $1 \leq k < \omega$. We say that r is *k-universal over* $(\mathcal{M}, q)$ iff whenever $A \in |\mathcal{M}|$ and $A \subseteq \rho_k^{\mathcal{M}}$, there is some term $\tau \in \mathrm{Sk}_k$ and $\bar{\alpha} \in \rho_k^{<\omega}$ such that

$$A = \tau^{\mathcal{M}}[\bar{\alpha}, r, q] \cap \rho_k^{\mathcal{M}} .$$

Again, we are interested in the case r is the kth standard parameter of $(\mathcal{M}, q)$. The k-universality of r will be used to show r remains the standard parameter in a certain hull of $\mathcal{M}$; the argument is given in the next lemma.

Lemma 2.8. *Let $\pi : \mathcal{H} \to \mathcal{M}$ be generalized $r\Sigma_k$ elementary , where $\mathcal{M}$ is a ppm and $1 \leq k < \omega$. Suppose $\rho_k^{\mathcal{M}} \subseteq \mathrm{OR}^{\mathcal{H}}$ and $\pi \restriction \rho_k^{\mathcal{M}} = $ identity. Suppose also that $\pi(r)$ is the kth standard parameter $(\mathcal{M}, \pi(q))$, and $\pi(r)$ is k-solid and k-universal over $(\mathcal{M}, \pi(q))$. Then $\rho_k^{\mathcal{H}} = \rho_k^{\mathcal{M}}$, r is the kth standard parameter of $(\mathcal{H}, q)$, and r is k-universal over $(\mathcal{H}, q)$.*

PROOF. For $\alpha \leq \rho_k^{\mathcal{M}}$ we have $\mathrm{Th}_k^{\mathcal{H}}(\alpha \cup \{s\}) = \mathrm{Th}_k^{\mathcal{M}}(\alpha \cup \{\pi(s)\})$, moreover the theory in question can be regarded as a subset of α. If $\alpha < \rho_k^{\mathcal{M}}$, then as $\rho_k^{\mathcal{M}}$ is a cardinal of $\mathcal{M}$ and $\mathcal{M}$ is strongly acceptable, $\mathrm{Th}_k^{\mathcal{M}}(\alpha \cup \{\pi(s)\})$ belongs to $\mathcal{H}$. This shows that $\rho_k^{\mathcal{M}} \leq \rho_k^{\mathcal{H}}$. But $\mathrm{Th}_k^{\mathcal{H}}(\rho_k^{\mathcal{M}} \cup \{\langle r, q \rangle\}) \notin |\mathcal{H}|$, as otherwise, letting $A \subseteq \rho_k^{\mathcal{M}}$ code it, we have $A = \pi(A) \cap \rho_k^{\mathcal{M}} \in |\mathcal{M}|$, and so $\mathrm{Th}_k^{\mathcal{M}}(\rho_k^{\mathcal{M}} \cup \{\pi(r), \pi(q)\}) \in |\mathcal{M}|$, a contradiction. Thus $\rho_k^{\mathcal{M}} = \rho_k^{\mathcal{H}}$.

We have $\mathrm{Th}_k^{\mathcal{H}}(\rho_k^{\mathcal{H}} \cup \{r, q\}) \notin |\mathcal{H}|$, so to see that r is the kth standard parameter of $(\mathcal{H}, q)$, suppose $s <_{\mathrm{lex}} r$. So $\pi(s) <_{\mathrm{lex}} \pi(r)$, and we have $A \subseteq \rho_k^{\mathcal{M}}$, $A \in |\mathcal{M}|$, such that A codes $\mathrm{Th}_k^{\mathcal{M}}(\rho_k^{\mathcal{M}} \cup \{\pi(s), \pi(q)\})$. The k universality of $\pi(r)$ over $(\mathcal{M}, \pi(q))$ easily implies $A \in \mathcal{H}$, and so $\mathrm{Th}_k^{\mathcal{H}}(\rho_k^{\mathcal{H}} \cup \{s, q\}) \in |\mathcal{H}|$, as desired.

It is routine to check that r is k-universal over $(\mathcal{H}, q)$. □

We now define by induction on $k \geq 0$

$$\mathfrak{C}_k(\mathcal{M}) = \text{the } k\text{th core of } \mathcal{M} ,$$
$$\rho_k(\mathcal{M}) = \text{the } k\text{th core projectum of } \mathcal{M} ,$$
$$p_k(\mathcal{M}) = \text{the } k\text{th core parameter of } M .$$

We shall assume that certain parameters we encounter in the course of the definition are solid and universal; otherwise we stop the induction. This assumption may not be necessary for a sensible definition, but it is true of the ppm we are interested in, as we shall show later.

DEFINITION 2.8.1. Let $\mathcal{M}$ be a ppm which is passive or active of types I or II. We define $\mathfrak{C}_k(\mathcal{M})$, $\rho_k(\mathcal{M})$, and $p_k(\mathcal{M})$ by induction on k.

$k = 0$: Let $\rho_0(\mathcal{M}) = \mathrm{OR}^{\mathcal{M}}$, $\mathfrak{C}_0(\mathcal{M}) = \mathcal{M}$, and $p_0(\mathcal{M}) = \varnothing$.

$k = 1$: Let $r =$ first standard parameter of $(\mathcal{M}, \varnothing)$. Suppose r is 1-universal over $(\mathcal{M}, \varnothing)$; otherwise stop the induction. Let

$$\pi : \mathcal{H}_1^{\mathcal{M}}(\rho_1^{\mathcal{M}} \cup \{r\}) \to \mathcal{M}$$

be the inverse of the collapse. If $\pi^{-1}(r)$ is not 1-solid over $(\mathcal{H}_1^{\mathcal{M}}(\rho_1^{\mathcal{M}} \cup \{r\}), \varnothing)$ then stop the induction, and otherwise set

$$\begin{aligned} \rho_1(\mathcal{M}) &= \rho_1^{\mathcal{M}} \\ p_1(\mathcal{M}) &= \langle r, \varnothing \rangle \\ \mathfrak{C}_1(\mathcal{M}) &= \mathcal{H}_1^{\mathcal{M}}(\rho_1^{\mathcal{M}} \cup \{r\}) . \end{aligned}$$

Notice that $p_1(\mathcal{M}) = \pi(\langle s, q \rangle)$, where s is the first standard parameter of $(\mathfrak{C}_1(\mathcal{M}), q)$, and s is 1-solid and 1-universal over $(\mathfrak{C}_1(\mathcal{M}), q)$. (This follows from Lemma 2.8.)

$k > 1$: Suppose we are given

$$p_{k-1}(\mathcal{M}) = \langle s, q \rangle$$

where s is the k-1st standard parameter of $(\mathfrak{C}_{k-1}(\mathcal{M}), q)$ and s is $k-1$ solid and $k-1$ universal over $(\mathfrak{C}_{k-1}(\mathcal{M}), q)$. Let

$$s = \langle \alpha_0 \cdots \alpha_\ell \rangle$$

and

$$b_i = \mathrm{Th}_{k-1}^{\mathfrak{C}_{k-1}(\mathcal{M})}(\alpha_i \cup \{\alpha_0, \ldots, \alpha_{i-1}, q\})$$

for $0 \le i \le \ell$, so that $b_i \in |\mathfrak{C}_{k-1}(\mathcal{M})|$ by solidity. Set

$$u_{k-1}(\mathcal{M}) = u = \begin{cases} \langle s, q, b_0 \cdots b_\ell \rangle & \text{if } \rho_{k-2}^{\mathfrak{C}_{k-1}(\mathcal{M})} = \mathrm{OR}^{\mathfrak{C}_{k-1}(\mathcal{M})} \\ \langle s, q, b_0 \cdots b_\ell, \rho_{k-2}^{\mathfrak{C}_{k-1}(\mathcal{M})} \rangle & \text{otherwise} . \end{cases}$$

Let r be the kth standard parameter of $(\mathfrak{C}_{k-1}(\mathcal{M}), u)$. If r is not k-solid and k-universal over $(\mathfrak{C}_{k-1}(\mathcal{M}), u)$, then stop the induction. If it is, consider

$$\pi : \mathcal{H}_k^{\mathfrak{C}_{k-1}(\mathcal{M})} \left(\rho_k^{\mathfrak{C}_{k-1}(m)} \cup \{r, u\} \right) \to \mathfrak{C}_{k-1}(\mathcal{M}) ,$$

the inverse of the collapse. Suppose that $\pi^{-1}(r)$ is k-solid over

$$\left(\mathcal{H}_k^{\mathfrak{C}_{k-1}(\mathcal{M})}\left(\rho_k^{\mathfrak{C}_{k-1}(\mathcal{M})}\cup\{r,u\}\right),\pi^{-1}(u)\right);$$

if not then we stop the induction. Set then

$$\rho_k(\mathcal{M}) = \rho_k^{\mathfrak{C}_{k-1}(\mathcal{M})},$$
$$p_k(\mathcal{M}) = \langle r, u\rangle,$$

and

$$\mathfrak{C}_k(\mathcal{M}) = \mathcal{H}_k^{\mathfrak{C}_{k-1}(\mathcal{M})}(\rho_k^{\mathfrak{C}_{k-1}(\mathcal{M})}\cup\{r,u\}).$$

DEFINITION 2.8.2. A ppm $\mathcal{M}$ *is k-solid* iff $\mathfrak{C}_k(\mathcal{M})$ is defined.

DEFINITION 2.8.3. $\mathcal{M}$ is *k-sound* iff $\mathcal{M}$ is k-solid and $\mathfrak{C}_i(\mathcal{M}) = \mathcal{M}$ for all $i \leq k$. $\mathcal{M}$ is a *core ppm* or: completely sound, or ω-sound, iff $\mathcal{M}$ is k sound for all $k < \omega$ such that $\rho_{k-1}^{\mathcal{M}} \neq 0$.

All levels of the model we eventually construct will be ω sound (for active type III levels this will be given a meaning in the next section). Nevertheless, one must consider ppm which are not ω-sound, as iteration of a core ppm can produce them.

Notice that if $\mathcal{M}$ is k-sound, then $\rho_i(\mathcal{M}) = \rho_i^{\mathcal{M}}$ for $i \leq k+1$. If $\mathcal{M}$ is not $i-1$ sound, then $\rho_i(\mathcal{M}) \neq \rho_i^{\mathcal{M}}$ is possible.

We record in a definition some properties of the natural embedding π mapping $\mathfrak{C}_{k+1}(\mathcal{M})$ into $\mathfrak{C}_k(\mathcal{M})$.

DEFINITION 2.8.4. Let $\pi : \mathcal{M} \to \mathcal{N}$, and $k \leq \omega$. We call π a *k-embedding* iff

(a) $\mathcal{M}$ and $\mathcal{N}$ are k-sound, (b) π is $r\Sigma_{k+1}$ elementary, (c) $\pi(p_i(\mathcal{M})) = p_i(\mathcal{N})$ for all $i \leq k$, (d) $\pi(\rho_i(\mathcal{M})) = \rho_i(\mathcal{N})$ for all $i \leq k-1$, and $\rho_k(\mathcal{N}) = \sup \pi''\rho_k(\mathcal{M})$.

(We adopt the convention that $\pi(\mathrm{OR}^{\mathcal{M}}) = \mathrm{OR}^{\mathcal{N}}$ in the previous definition.)

Lemma 2.9. *Let $\mathcal{M}$ be a $k+1$ solid ppm, and $\pi : \mathfrak{C}_{k+1}(\mathcal{M}) \to \mathfrak{C}_k(\mathcal{M})$ be the inverse of the collapse. Then π is a k-embedding ; moreover $\mathfrak{C}_{k+1}(\mathcal{M})$ is $k+1$ sound and $\pi(\rho_{k+1}(\mathfrak{C}_{k+1(\mathcal{M})})) = \rho_{k+1}(\mathfrak{C}_k(\mathcal{M})$.*

Remark. It is easy to see that if $\pi : \mathcal{M} \to \mathcal{N}$ is a k-embedding, then $\pi(u_k(\mathcal{M})) = u_k(\mathcal{N})$.

Appendix to §2

We close this section by relating the $r\Sigma_n$ hierarchy to the more traditional hierarchy involving master codes and iterated Σ_1 definability.

First, the use of generalized $r\Sigma_{n+1}$ formulae rather than just pure $r\Sigma_{n+1}$ formulae in defining $\mathrm{Th}_{n+1}^{\mathcal{M}}(X)$ does not affect the value of $\rho_{n+1}^{\mathcal{M}}$, at least if $\mathcal{M}$ is n sound.

Lemma 2.10. *Let $\mathcal{M}$ be a ppm and $n \geq 0$. Let $q \in |\mathcal{M}|$, and suppose in the case that $n \geq 1$ that $\mathcal{M} = \mathcal{H}_n^{\mathcal{M}}(\rho_n^{\mathcal{M}} \cup \{q\})$. Suppose that*

$$\mathrm{Th}_{n+1}^{\mathcal{M}}(\alpha \cup \{q\}) \cap \{\langle \varphi, \bar{a} \rangle \mid \varphi \textit{ is pure } r\Sigma_{n+1}\}$$

is a member of $|\mathcal{M}|$. Then in fact $\mathrm{Th}_{n+1}^{\mathcal{M}}(\alpha \cup \{q\}) \in |\mathcal{M}|$.

PROOF. We give the proof in full only for $n = 0$ and $\mathcal{M}$ passive and $\mathrm{OR}^{\mathcal{M}} = \omega\lambda$, for λ a limit. So let us make those assumptions. Let

$$\mathcal{M} = (J_\lambda^{\vec{E}}, \in, \vec{E}) \qquad (\lambda \quad \text{limit})$$

and

$$\mathcal{M}_\beta = (J_\beta^{\vec{E}}, \in, \vec{E} \restriction \beta)$$

for $\beta < \lambda$. For $\tau \in \mathrm{Sk}_1$, $\beta < \lambda$, and $\bar{u} \in |\mathcal{M}_\beta|^{<\omega}$, we say that $\tau(\bar{u})$ changes value at β iff

$$\tau^{\mathcal{M}_\beta}[\bar{u}] \neq \tau^{\mathcal{M}_{\beta+1}}[\bar{u}].$$

Notice that if $\tau_\varphi \in \mathrm{Sk}_1$ is a basic term, then $\tau_\varphi(\bar{u})$ changes value finitely often, since the new value precedes the old in the order of construction (unless the old is 0). It follows that for any $\tau \in \mathrm{Sk}_1$, $\tau(\bar{u})$ changes value finitely often. Notice also that there is a recursive map $(\tau, n) \mapsto \theta_{\tau,n}$ associating to each $\tau \in \mathrm{Sk}_1$ and $n < \omega$ an $r\Sigma_1$ formula $\theta_{\tau,n}$ such that $\mathcal{M} \models \theta_{\tau,n}[\bar{u}, a]$ if and only if $\tau(\bar{u})$ changes value at least n times and a is the nth value of $\tau(\bar{u})$. Now, letting

$$P = \mathrm{Th}_1^{\mathcal{M}}(\alpha \cup \{q\}) \cap \{\langle \varphi, \bar{a} \rangle \mid \varphi \text{ is pure } r\Sigma_{n+1}\}$$

we can compute $\mathrm{Th}_1^{\mathcal{M}}(\alpha \cup \{q\})$ from P inside $\mathcal{M}$ as follows: Given a potential member $\langle \varphi, \bar{a} \rangle$ of $\mathrm{Th}_1^{\mathcal{M}}(\alpha \cup \{q\})$, which we write as $\psi(\tau_1(\bar{u}) \cdots \tau_k(\bar{u}))$ where ψ is pure $r\Sigma_1$ and $\tau_1 \cdots \tau_k \in \mathrm{Sk}_1$ and $\bar{u} \in (\alpha \cup \{q\})^{<\omega}$, find first numbers $n_1 \cdots n_k < \omega$ such that

$$\bigwedge_{i \leq k} \exists a \, \theta_{\tau_{i,n_i}}(\bar{u}, a) \in P$$

and

$$\bigvee_{i \leq k} \exists a \, \theta_{\tau_{i,n_i+1}}(\bar{u}, a) \notin P.$$

Then $\langle \varphi, \bar{a} \rangle \in \mathrm{Th}_1^{\mathcal{M}}(\alpha \cup \{q\}$ iff

$$\exists a_1 \cdots \exists a_k \left(\bigwedge_{i \leq k} \theta_{\tau_i, n_i}(\bar{u}, a_i) \wedge \psi(a_1 \cdots a_k) \right)$$

is a member of P.

If $M = (J_{\alpha+1}^{\vec{E}}, \in, \vec{E})$, then we can use the $S_{\omega\alpha+n}^{\vec{E}}$, for $n < \omega$, as we used the M_β's of the previous argument.

If M is active, then we can use the fact that $r\Sigma_1^m = \Sigma_1$ over M^*, where M^* is the amenable structure associated to M (cf. the remark following Corollary 2.2.) We can then ramify Σ_1 over M^* as above to carry out the proof.

This finishes the case $n = 0$. Now let $n \geq 1$. If $\tau \in \mathrm{Sk}_n$ and $\bar{\beta} \in (\rho_n^{\mathcal{M}})^{<\omega}$, then we call the triple $(\tau, \bar{\beta}, q)$ a *name* of $\tau^{\mathcal{M}}[\bar{\beta}, q]$. We are assuming every member of $|\mathcal{M}|$ has a name. If θ is $r\Sigma_{n+1}$ and $a_1 \cdots a_k$ are names and $\beta < \rho_n^{\mathcal{M}}$, then we let "$\mathrm{Th}_n^{\mathcal{M}}(\beta \cup \{q\})$ witnesses that $\theta(a_1 \cdots a_k)$" have the obvious meaning. Namely, if

$$\theta = \exists a, b(\dot{T}_n(a, b) \wedge \psi(a, b, v_1 \cdots v_k))$$

and

$$a_i = (\tau_i, \bar{\beta}_i, q),$$

then $\mathrm{Th}_n^{\mathcal{M}}(\beta \cup \{q\})$ witnesses $\theta(a_1 \cdots a_k)$ iff there is a $\gamma < \beta$ and there are names $(\sigma_1, \bar{\eta}_1, q)$ and $(\sigma_2, \bar{\eta}_2, q)$ such that, first, the following sentences are in $\mathrm{Th}_n^{\mathcal{M}}(\beta \cup \{q\})$:

(a) $\sigma_1(\bar{\eta}_1, q) = \langle \gamma, x \rangle$ for some x such that $\sigma_2(\bar{\eta}_2, q)$ is a complete generalized $r\Sigma_n$ theory of parameters in $\gamma \cup \{x\}$.

(b) $\psi(\sigma_1(\bar{\eta}_1, q), \sigma_2(\bar{\eta}_2, q), \tau_1(\bar{\beta}_1, q) \cdots \tau_k(\bar{\beta}_k, q))$.

(implicit here is that $\bar{\beta}_i, \bar{\eta}_i \in \beta^{<\omega}$), and second, if we let $\sigma_1^{*\mathcal{M}}[\bar{\eta}_1, q]$ = second coordinate of $\sigma_1^{\mathcal{M}}[\bar{\eta}_1, q]$, and we let $f(\varphi, \bar{\delta})$ = a canonical name for $(\varphi, \bar{\delta}^\frown \sigma_1^{*\mathcal{M}}[\bar{\eta}_1, q])$ for each generalized $r\Sigma_n$ formula φ and $\bar{\delta} \in \gamma^{<\omega}$, then

$$\text{“}f(\varphi, \bar{\delta}) \in \sigma_2(\bar{\eta}_2, q)\text{”} \in \mathrm{Th}_n^{\mathcal{M}}(\beta \cup \{q\}) \text{iff “}\varphi(\bar{\delta}, \sigma^*(\bar{\eta}_1, q))\text{”} \in \mathrm{Th}_n^{\mathcal{M}}(\beta \cup \{q\})$$

Remark. We have taken some liberties above, as $\mathrm{Th}_n^{\mathcal{M}}(\beta \cup \{q\})$ is not literally speaking a set of sentences.

Now for $\tau \in \mathrm{Sk}_{n+1}$ and $a_1 \cdots a_k$, b names and $\beta < \rho_n^{\mathcal{M}}$ let

$$\tau^\beta(a_1 \cdots a_k) = b \quad \text{iff} \quad \mathrm{Th}_n^{\mathcal{M}}(\beta \cup \{q\}) \text{ witnesses } \tau(a_1 \cdots a_k) = b$$

where the right hand side is interpreted in the spirit above. We can use the τ^β's to carry out the argument given in the case $n = 0$. □

The calculations just indicated also give

Lemma 2.11. *Let $\mathcal{M}$ be a ppm, $q \in |\mathcal{M}|$, and $\mathcal{M} = \mathcal{H}_n^{\mathcal{M}}(\rho_n^{\mathcal{M}} \cup \{q\})$ where $n \geq 1$. Let*

$$A = \mathrm{Th}_n^{\mathcal{M}}(\rho_n^{\mathcal{M}} \cup \{q\}) \cap \{(\varphi, \bar{c}) \mid \varphi \quad \textit{is pure } r\Sigma_n\},$$

coded in a natural way as a subset of $\rho_n^{\mathcal{M}}$. Then, letting

$$\mathcal{N} = (J_{\rho_n^{\mathcal{M}}}^{\dot{E}^{\mathcal{M}}}, \in, \dot{E}^{\mathcal{M}} \restriction \rho_n^{\mathcal{M}}, A)$$

$\mathcal{N}$ is amenable, and for all $B \subseteq \rho_n^{\mathcal{M}}$, B is $r\Sigma_{n+1}^{\mathcal{M}}$ iff B is Σ_1 over $\mathcal{N}$.

PROOF. From $A \cap \alpha$ we can compute $\mathrm{Th}_n^{\mathcal{M}}(\alpha \cup \{q\})$ in a simple way, as in the preceding lemma. Thus we may as well assume $A = \mathrm{Th}_n^{\mathcal{M}}(\rho_n^{\mathcal{M}} \cup \{q\})$. Now suppose

$$\eta \in B \Leftrightarrow \mathcal{M} \models \varphi[\eta, x]$$

and let $x = \sigma[\bar{\beta}, q]$, $\sigma \in \mathrm{Sk}_n$, $\bar{\beta} \in (\rho_n^{\mathcal{M}})^{<\omega}$, where φ is $r\Sigma_{n+1}$. Then

$$\eta \in B \Leftrightarrow \exists \beta < \rho_n^{\mathcal{M}}(\mathrm{Th}_n^{\mathcal{M}}(\beta \cup \{q\}) \quad \text{witnesses} \quad \varphi(\eta^*, \sigma(\bar{\beta}, q)))$$

where η^* is a canonical name for η. This shows B is Σ_1 over $\mathcal{N}$. The converse is easy. □

§3. Squashed Mice

Let $\mathcal{M}$ be an active type III ppm. Let E be an extender over $\mathcal{M}$ with $\kappa = \operatorname{crit} E < \nu^{\mathcal{M}}$. Even if wellfounded, $\operatorname{Ult}(\mathcal{M}, E)$ may not be a ppm. The trouble is in the initial segment condition: if $i_E''\nu^{\mathcal{M}}$ is not cofinal in $i_E(\nu^{\mathcal{M}})$, then this condition will fail in $\operatorname{Ult}(\mathcal{M}, E)$. The problem seems to be that we are using too many functions in forming $\operatorname{Ult}(\mathcal{M}, E)$; we'd like to use only functions in $J_{\nu^{\mathcal{M}}}^{\dot{E}^{\mathcal{M}}}$ in order to get continuity of i_E at $\nu^{\mathcal{M}}$. Lemma 9.1 and the remarks following it give a fuller explanation. This leads to

DEFINITION 3.0.1. ($\mathcal{M}$-squash) Let $\mathcal{M}$ be an active type III ppm. Let F be the extender coded by $\dot{F}^{\mathcal{M}}$ and $\nu = \nu^{\mathcal{M}}$. Then

$$\mathcal{M}^{sq} = (J_{\nu}^{\dot{E}^{\mathcal{M}}}, \in, \dot{E}^{\mathcal{M}} \restriction \nu, F \restriction \nu)$$

The symbol $\mathcal{M}^{sq}$ stands for "$\mathcal{M}$-squash". The term "squashed mouse" was invented by Dodd for use in a similar, but more complicated, context.

Recall that $\nu^{\mathcal{M}}$ is a cardinal of $\mathcal{M}$ in the type III case, so that $\mathcal{M}^{sq}$ includes all sets which have hereditarily cardinality $< \nu^{\mathcal{M}}$ in $\mathcal{M}$. Our next lemma shows that $\mathcal{M}^{sq}$ is amenable.

Lemma 3.1. *Let $\mathcal{M}$ be an active type III ppm. Then there are cofinally many $\gamma < \nu^{\mathcal{M}}$ such that $\dot{E}_{\gamma}^{m} = F \restriction \gamma$ where F is the extender coded by $\dot{F}^{\mathcal{M}}$.*

PROOF. Let $\kappa = \operatorname{crit} F$, and let $\eta - 1$ be a generator of F. By the initial segment condition, there is a $\gamma < \operatorname{OR}^{\mathcal{M}}$ such that $\dot{E}_{\gamma}^{\mathcal{M}}$ exists and is the trivial completion of $F \restriction \eta$. (Alternative (b) of the initial segment condition cannot hold as η is a successor ordinal.) Now the natural map π from $\operatorname{Ult}(\mathcal{M}, F \restriction \eta)$ into $\operatorname{Ult}(\mathcal{M}, F)$ has critical point $\geq \eta$, and hence $\operatorname{crit}(\pi) \geq \gamma$ since $\gamma = (\eta^+)^{\operatorname{Ult}(\mathcal{M}, F \restriction \eta)}$. This implies that $F \restriction \gamma$ is the trivial completion of $F \restriction \eta$, which is $\dot{E}_{\gamma}^{\mathcal{M}}$.

To see this let G be the trivial completion of $F \restriction \eta$. We have

$$\begin{array}{ccc} M & \xrightarrow{i_F} & \operatorname{Ult}(M, F) \\ {\scriptstyle i_G} \searrow & & \uparrow {\scriptstyle \pi} \\ & \operatorname{Ult}(M, F \restriction \eta) & \end{array}$$

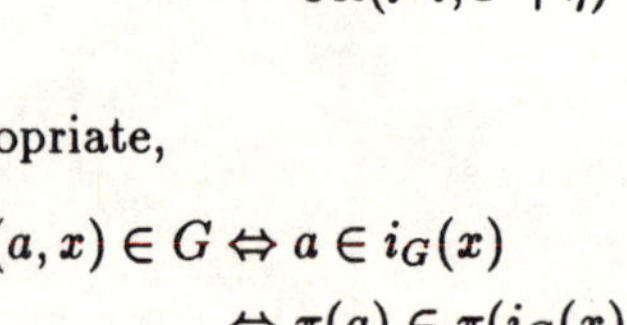

and for $a \in [\gamma]^{<\omega}$, x appropriate,

$$(a, x) \in G \Leftrightarrow a \in i_G(x)$$
$$\Leftrightarrow \pi(a) \in \pi(i_G(x))$$
$$\Leftrightarrow a \in i_F(x)$$
$$\Leftrightarrow (a, x) \in F.$$

Since there are arbitrarily large $\eta < \nu^{\mathcal{M}}$ such that $\eta - 1$ is a generator of F, this completes the proof. □

So $\mathcal{M}^{sq}$ is amenable. Moreover, the definition of $\nu^{\mathcal{M}}$ guarantees that the rest of $\mathcal{M}$ can be recovered from $\mathcal{M}^{sq}$ by taking an ultrapower.

If E is an extender over $\mathcal{M}$ with crit $E < \nu^{\mathcal{M}}$, we'll have

$$\begin{array}{ccc} \mathcal{M} & \longrightarrow & \mathrm{Ult}(\mathcal{M}, E) \\ {\scriptstyle \mathrm{id}}\uparrow & & \uparrow \\ \mathcal{M}^{sq} & \longrightarrow & \mathrm{Ult}(\mathcal{M}^{sq}, E) \end{array}$$

and $\mathrm{Ult}(\mathcal{M}^{sq}, E) = \mathcal{N}^{sq}$ for some $\mathcal{N} \subseteq \mathrm{Ult}(\mathcal{M}, E)$. But $\mathcal{N} \neq \mathrm{Ult}(\mathcal{M}, E)$ is possible, and this is what leads us to iterate on the squashed level.

As we shall iterate $\mathcal{M}^{sq}$ and not $\mathcal{M}$, the appropriate definability hierarchy is based on $\mathcal{M}^{sq}$, not $\mathcal{M}$ as in §2. Note every $\mathcal{M}$-definable subset of $\nu^{\mathcal{M}}$ is definable over $\mathcal{M}^{sq}$.

Definition 3.1.1. $\mathcal{N}$ is an sppm iff $\mathcal{N} = \mathcal{M}^{sq}$ for some active type III ppm $\mathcal{M}$.

We now introduce a language appropriate for sppm.

Definition 3.1.2. $\mathcal{L}^*$ is the language of set theory with additional 1- place predicate symbol $\dot{E}$, 2-place predicate symbol $\dot{F}$, and constant symbol $\dot{\mu}$.

We interpret $\mathcal{L}^*$ in an sppm

$$\mathcal{N} = (J_\nu^{\vec{E}}, \in, \vec{E}, F)$$

by setting $\dot{E}^{\mathcal{N}} = \vec{E}, \dot{F}^{\mathcal{N}} = F$, and $\dot{\mu}^{\mathcal{N}} = \mathrm{crit}\ F$.

As sppm are amenable with respect to their predicates, we can work with the usual notions of Σ_0 and Σ_1.

Definition 3.1.3. (a) A formula of $\mathcal{L}^*$ is Σ_0 iff it is built up from atomic formulae using $\wedge, \vee, \neg, \exists x \in y$, and $\forall x \in y$.

(b) The Σ_n and $\prod_n$ formulae of $\mathcal{L}^*$ are also as usual.

We want now to say "I am an sppm" with a simple formula.

Definition 3.1.4. A *P-formula* is a formula of $\mathcal{L}^*$ of the form

$$\theta(\bar{v}) = \forall x \exists y (x \subset y \wedge \psi(y) \wedge \forall a \in x \, \exists b \in y \, \varphi(a, b, \bar{v}),$$

where ψ is Σ_1 without x free in it, and φ is Σ_0 without x or y free in it.

Thus a P formula can say a little more than that there are cofinally many y (under $\subseteq$) with a Σ_1 property. We aren't sure how necessary the little more is, but as the preservation lemma still goes through, there's no harm in it.

Lemma 3.2. *Let $\mathcal{M}$ and $\mathcal{N}$ be transitive $\mathcal{L}^*$ structures, and $\pi : \mathcal{M} \to \mathcal{N}$, and ψ be a P formula.*

(a) *If π is a Σ_1 embedding and $\mathcal{N} \models \psi[\pi(\bar{a})]$, then $\mathcal{M} \models \psi[\bar{a}]$.*

(b) *If π is a cofinal (i.e. $|\mathcal{N}| = \bigcup \operatorname{ran} \pi$) Σ_0 embedding and $\mathcal{M} \models \psi[\bar{a}]$, then $\mathcal{N} \models \psi[\pi(\bar{a})]$.*

One can't quite say "I am an sppm" with a P sentence, since the decoding of $\mathcal{M}$ from $\mathcal{M}^{sq}$ requires taking an ultrapower, and we can't capture the wellfoundedness of this ultrapower. We do get

Lemma 3.3. *There is a P sentence ψ of $\mathcal{L}^*$ such that*

(a) *If $\mathcal{N}$ is an sppm, then $\mathcal{N} \models \psi$.*

(b) *If $\mathcal{N}$ is transitive and $\mathcal{N} \models \psi$, then $\dot{F}^{\mathcal{N}}$ is a pre-extender over $\mathcal{N}$; moreover, if $\operatorname{Ult}(\mathcal{N}, \dot{F}^{\mathcal{N}})$ is wellfounded then $\mathcal{N}$ is an sppm or $\mathcal{N}$ is "of superstrong type", that is $i_F^{\mathcal{N}}(\text{crit } F) = \text{length } F = OR^{\mathcal{N}}$).*

PROOF (Sketch). By Dodd-Jensen we have a P sentence θ_1 whose transitive models $\mathcal{N}$ are those of the form $\mathcal{N} = (J_\nu^{\dot{E}^{\mathcal{N}}}, \dots)$, ν a limit ordinal.

Let θ_2 be the Π_1 sentence of $\mathcal{L}^*$ asserting that $\dot{E}^{\mathcal{N}}$ is good at all $\alpha < \mathrm{OR}^{\mathcal{N}}$.

Let θ_3 be the Π_1 sentence: $\forall a \forall x(\dot{F}(a, x) \Rightarrow a \in [\mathrm{OR}]^{<\omega} \wedge x \subseteq [\dot{\mu}]^{\operatorname{card} a})$

Let θ_4 be the P sentence: There are cofinally many ordinals γ such that $\gamma \in \operatorname{dom} \dot{E}$ and $J_\gamma^{\vec{E}} \cap E_\gamma = \dot{F} \restriction \gamma \cap J_\gamma^{\vec{E}}$.

It may seem that "$\dot{\mu}^+$ exists" is Σ_2, but we can say with θ_5:

$$\exists \text{ ordinal } \alpha \text{ such that } \dot{\mu} < \alpha \text{ and } \{(\beta_0, \beta_1) \mid J_{\dot{\mu}}^{\dot{E}} \models \beta_1 = \beta_0^+\} \in \dot{F}_{\{\dot{\mu}, \alpha\}} .$$

We claim $\psi = \bigwedge_{i \leq 5} \theta_i$ is as desired. Clearly, if $\mathcal{N}$ is an sppm, then $\mathcal{N} \models \bigwedge_{i \leq 5} \theta_i$.

Now suppose $\mathcal{N}$ is a transitive $\mathcal{L}^*$ structure such that $\mathcal{N} \models \bigwedge_{i \leq 5} \theta_i$. As $\mathcal{N} \models \theta_4$, we see that $\dot{F}^{\mathcal{N}} = F$ is a pre-extender over $\mathcal{N}$. Suppose that $\operatorname{Ult}(\mathcal{N}, F)$ is wellfounded, and that $i_F^{\mathcal{N}}(\text{crit } F) > \mathrm{OR}^{\mathcal{N}}$. Let

$$\nu = \mathrm{OR}^{\mathcal{N}}$$
$$\alpha = (\nu^+)^{\operatorname{Ult}(\mathcal{N}, F)}$$
$$G = \text{the } (\dot{\mu}^{\mathcal{N}}, \alpha) \text{ extender derived from } i_F^{\mathcal{N}} : \mathcal{N} \to \operatorname{Ult}(\mathcal{N}, F)$$
$$\mathcal{M} = \left(J_\alpha^{i_F(\dot{E}^{\mathcal{N}})}, \in, i_F(\dot{E}^{\mathcal{N}}) \restriction \alpha, G\right) .$$

Note that α exists since $i_F(\dot\mu^{\mathcal{N}}) > \nu$.

We claim $\mathcal{M}$ is an active type III ppm, and $\nu = \nu^{\mathcal{M}}$. For this, note $i_F(\dot E^{\mathcal{N}}) \restriction \alpha$ is good at all $\beta < \alpha$, since $\dot E^{\mathcal{N}}$ is good at all $\beta < \dot\mu^{\mathcal{N}}$. So it is enough to check $i_F(\dot E^{\mathcal{N}}) \restriction \alpha^\frown G$ is good at α.

Clearly $\mathcal{M}$ is strongly acceptable. G is a pre-extender over $\mathcal{M}$ as G is a pre-extender over $\mathcal{N}$ and

$$P(\dot\mu^{\mathcal{N}}) \cap \mathcal{N} = P(\dot\mu^{\mathcal{N}}) \cap \mathrm{Ult}(\mathcal{N}, F) = P(\dot\mu^{\mathcal{N}}) \cap \mathcal{M}\,.$$

The ordinal ν satisfies condition 3 of good at α since $P(\dot\mu^{\mathcal{N}}) \cap |\mathcal{M}| \subseteq J_\nu^{i_F(\dot E^{\mathcal{N}})}$ as $\mathcal{N} \models \theta_5$. Since G is derived from $i_F^{\mathcal{N}}$, $\mathrm{Ult}(\mathcal{N}, G) = \mathrm{Ult}(\mathcal{N}, F)$; on the other hand $\mathcal{N}$ and $\mathcal{M}$ agree up to ν (i.e. $J_\nu^{i_F(\dot E^{\mathcal{N}})} = J_\nu^{\dot E^{\mathcal{N}}}$) as $\mathcal{N} \models \theta_4$ so $\mathrm{Ult}(\mathcal{M}, G)$ agrees with $\mathrm{Ult}(\mathcal{N}, G)$ up to $i_F^{\mathcal{N}}(\nu) = i_G^{\mathcal{M}}(\nu)$, so that $\alpha = \nu^+$ in $\mathrm{Ult}(\mathcal{M}, G)$. This verifies 3(a). For 3(b), note $G \restriction \nu = F$, and that if $\beta < \alpha$ then for some $a \subseteq \nu$ and $f : [\dot\mu^{\mathcal{N}}]^{\operatorname{card} a} \to \dot\mu^{\mathcal{N}}$ such that $f \in \mathcal{N}$ we have

$$[a, f]_F^{\mathcal{N}} = \beta\,,$$

so

$$[a, f]_G^{\mathcal{N}} = \beta\,,$$

so

$$[a, f]_G^{\mathcal{M}} = \beta\,.$$

This is enough to give 3(b). Finally, ν is the least ordinals satisfying clause 3 since if $\gamma < \nu$, then $G \restriction \gamma = F \restriction \gamma \in \mathcal{M}$ by the fact that $\mathcal{N} \models \theta_4$.

It is easy to see the coherence condition 4 is satisfied. The initial segment condition (only 5(a) is relevant) is satisfied as $\mathcal{N} \models \theta_4$ and $i_F^{\mathcal{N}}(\dot E^{\mathcal{N}}) \restriction \nu = \dot E^{\mathcal{N}}$.

Thus $\mathcal{M}$ is an active type III ppm with $\nu = \nu^{\mathcal{M}}$. Clearly $\mathcal{N} = \mathcal{M}^{sq}$. □

Remark. It is annoying that we must include the possibility that $\mathcal{N} \models \psi$ be "of superstrong type", but our attempts to strengthen ψ so as to exclude this have not succeeded. Notice that if $\mathcal{N} \models \psi$ is of superstrong type, then a standard argument gives

$$(J_\nu^{\dot E^{\mathcal{N}}}, \in, \dot E^{\mathcal{N}}) \models \mathrm{ZFC} + \dot\mu^{\mathcal{N}} \text{ is a Shelah limit of Shelah cardinals}\,.$$

($\nu = \mathrm{OR}^{\mathcal{N}}$). So $\mathcal{N}$ is far above any mice our theory can handle anyway.

The rest of this section is an obvious parallel to §2. Because sppm are amenable, we could adopt a very literal version of the Dodd-Jensen approach here (in particular, we could stick to the usual Σ_n hierarchy); however, for the sake of internal consistency, we shall adopt the approach of §2.

Skolem terms and projecta.

DEFINITION 3.3.1. $\mathcal{L}^{**}$ is $\mathcal{L}^*$ together with binary relation symbols $\dot{T}_n$ for $1 \leq n < \omega$.

We define the quasi-Σ_n formulae for $n \geq 1$.

DEFINITION 3.3.2.

(a) The $q\Sigma_1$ formulae of $\mathcal{L}^{**}$ are precisely the Σ_1 formulae of $\mathcal{L}^*$.
(b) A formula $\theta(\bar{v})$ of $\mathcal{L}^{**}$ is $q\Sigma_{n+1}$, where $n \geq 1$, iff

$$\theta(\bar{v}) = \exists a \exists b(\dot{T}_n(a,b) \wedge \varphi(a,b,\bar{v}))$$

where φ is $q\Sigma_1$.

DEFINITION 3.3.3. For $\varphi(v_0 \cdots v_{k+1})$ an $\mathcal{L}^{**}$ formulae, $\tau_\varphi(v_0 \cdots v_k)$ is the basic Skolem term associated to φ. Having interpreted φ in an sppm $\mathcal{N}$, we set

$$\tau_\rho^{\mathcal{N}}[a_0 \cdots a_k] = \begin{cases} <^{\mathcal{N}} & \text{least } b \text{ such that } \mathcal{N} \models \varphi[\bar{a}, b] \\ 0 & \text{if no such } b \text{ exists}. \end{cases}$$

DEFINITION 3.3.4. SK_n (for $n \geq 1$) is the smallest class of terms containing all τ_φ for φ $q\Sigma_n$ and closed under composition.

DEFINITION 3.3.5. A $\mathcal{L}^{**}$ formula is *generalized* $q\Sigma_n$ iff it results from substituting terms in SK_n for free variables of a $q\Sigma_n$ formula. (The substitution must be such that no free variable of a term becomes bound in the result.)

DEFINITION 3.3.6. Let $\mathcal{M}$ be an sppm. Then for $n \geq 1$

(a) $\mathrm{Th}_n^{\mathcal{M}}(X) = \{(\varphi, \bar{a}) \mid \varphi \text{ is generalized } q\Sigma_n \text{ and } \bar{a} \in X^{<\omega} \text{ and } \mathcal{M} \models \varphi[\bar{a}]\}$.
(b) $\rho_n^{\mathcal{M}} =$ least $\alpha \leq \mathrm{OR}^{\mathcal{M}}$ such that for some $p \in |\mathcal{M}|$, $\mathrm{Th}_n^{\mathcal{M}}(\alpha \cup \{p\}) \notin |\mathcal{M}|$.
(c) $\dot{T}_n^{\mathcal{M}}(a,b)$ iff $a = \langle \alpha, q \rangle$ for some $\alpha < \rho_n^{\mathcal{M}}$ such that $b = \mathrm{Th}_n^{\mathcal{M}}(\alpha \cup \{q\})$.

We define the classes of relations $q\Sigma_n^{\mathcal{M}}$, etc., in the obvious way. It is easy to see that $q\Sigma_n^{\mathcal{M}}$ is closed under $\exists$, $\wedge$, and $\vee$, and that $(q\Sigma_n^{\mathcal{M}} \cup q\prod_n^{\mathcal{M}}) \subseteq q\Sigma_{n+1}^{\mathcal{M}}$, uniformly over all sppm. One can also show $\neg\dot{T}_n^{\mathcal{M}}$ is a $q\Sigma_{n+1}^m$ relation (uniformly) in parameter $\rho_n^{\mathcal{M}}$.

Hulls.

For $\mathcal{M}$ a sppm and $X \subseteq |\mathcal{M}|$ and $n \geq 1$ let

$$H_n^{\mathcal{M}}(X) = \text{the transitive collapse of } \{\tau^{\mathcal{M}}[\bar{a}] \mid \bar{a} \in X^{<\omega} \text{ and } \tau \in \mathrm{SK}_n\}$$
$$\mathcal{H}_n^{\mathcal{M}}(X) = (H_n^{\mathcal{M}}(X), \in, \pi''(\dot{E}^{\mathcal{M}}), \pi''(\dot{F}^{\mathcal{M}}))$$

where π is the collapse map.

Lemma 3.4. *Let $\mathcal{M}$ be an sppm such that $\forall\nu \in \mathrm{OR}^{\mathcal{M}}$ $(J_\nu^{\dot{E}^{\mathcal{M}}} \models$ there are no Shelah cardinals). Then for any $X \subseteq |\mathcal{M}|$ and $n \geq 1$, $\mathcal{H}_n^{\mathcal{M}}(X)$ is an sppm.*

PROOF. By Lemmas 3.2 and 3.3. Strictly speaking, we haven't packed enough wellfoundedness of ultrapowers into being an sppm to be able just to quote 3.3(b), but the proof of 3.3(b) requires only the wellfoundedness we have.

Lemma 2.7 carries over verbatim.

Lemma 3.5. *Lemma* 2.7 *remains true if one replaces "ppm which is passive or active of types* I *or* II*" by "sppm" and "$r\Sigma_n$" by "$q\Sigma_n$".*

Standard parameters, solid parameters, and Cores.

The definitions and results of §2 carry over verbatim. (The only "results" here are Lemmas 2.8, 2.9.) We shall say no more.

Premice.

DEFINITION 3.5.1. Let $\mathcal{M} = \mathcal{J}_\alpha^{\vec{E}}$ be a ppm. We say $\mathcal{M}$ is a premouse iff for all $\beta < \alpha$,

(1) $\mathcal{J}_\beta^{\vec{E}}$ is passive or active of types I or II $\Rightarrow \mathcal{J}_\beta^{\vec{E}}$ is ω-sound, and

(2) $\mathcal{J}_\beta^{\vec{E}}$ is active type III $\Rightarrow (\mathcal{J}_\beta^{\vec{E}})^{\mathrm{sq}}$ is ω-sound.

Notice that a premouse need not itself be ω-sound.

We shall eventually build an $\vec{E}$ such that every $\mathcal{J}_\alpha^{\vec{E}}$ is a premouse.

§4. ULTRAPOWERS

Let $\mathcal{M}$ be either a ppm or an sppm, and $\kappa < \rho_n^{\mathcal{M}}$. Let E be a (κ, λ) pre-extender over $\mathcal{M}$. (We are interested in the case that for some ppm or sppm $\mathcal{N}$ such that $P(\kappa)^{\mathcal{N}} = P(\kappa)^{M}$ we have $E = \dot{F}^{\mathcal{N}}$ or $\exists\gamma(E = \dot{E}_\gamma^{\mathcal{N}})$. It is easy to check that E is a pre-extender over $\mathcal{M}$ in this case.) We wish to define $\mathrm{Ult}_n(\mathcal{M}, E)$.

We begin with the universe of $\mathrm{Ult}_n(\mathcal{M}, E)$ and the $\in$ relation on it.

If $n = 0$, then the elements of $\mathrm{Ult}_0(\mathcal{M}, E)$ are equivalence classes $[a, f]_E^{\mathcal{M}}$, where $a \subseteq \lambda$ is finite and $f \in |\mathcal{M}|$ has domain $[\kappa]^{\mathrm{card}\, a}$. The equivalence relation is as usual: $(a, f) \sim (b, g)$ iff for $E_{a\cup b}$ a.e. $\bar{x}$, $\tilde{f}(\bar{x}) = \tilde{g}(\bar{x})$ where $\tilde{f}$ and $\tilde{g}$ come from f and g by adding the appropriate dummy variables. E measures enough sets that the definition makes sense. The $\in$ relation on equivalence classes is as usual.

If $n > 0$, then let $\tau = \tau(v_0 \cdots v_i)$ be a term in Sk_n if $\mathcal{M}$ is a ppm or in SK_n if $\mathcal{M}$ is an sppm. Let $q \in |\mathcal{M}|$. Then for $\bar{\alpha} \in [\kappa]^i$

$$f_{\tau,q}(\bar{\alpha}) = \tau^{\mathcal{M}}[\bar{\alpha}, q]\,.$$

The elements of $\mathrm{Ult}_n(\mathcal{M}, E)$ are equivalence classes $[a, f]_E^{\mathcal{M}}$ where $a \subseteq \lambda$ is finite and $f = f_{\tau,q}$ for some $q \in |\mathcal{M}|$ and $\tau \in \mathrm{Sk}_n$ (resp. SK_n). The equivalence relation is as usual. E measures enough sets that the definition makes sense because $\kappa < \rho_n^{\mathcal{M}}$. Again, the $\in$ relation is as usual.

$\mathrm{Ult}_n(\mathcal{M}, E)$ may be illfounded; however, if it is wellfounded we shall identify it with the transitive set to which it is isomorphic.

We must define $\dot{E}^{\mathrm{Ult}_n(\mathcal{M},E)}$ and $\dot{F}^{\mathrm{Ult}_n(\mathcal{M},E)}$ to complete the definition of the structure $\mathrm{Ult}_n(\mathcal{M}, E)$. Let

$$[\langle a, f\rangle]_E^{\mathcal{M}} \in \dot{E}^{\mathrm{Ult}_n(\mathcal{M},E)} \text{ iff } \{\,\bar{\alpha} \mid f(\bar{\alpha}) \in \dot{E}^{\mathcal{M}}\,\} \in E_a\,.$$

It is easy to see that E_a measures the set in question, using the amenability of $\mathcal{M}$ with resp. to $\dot{E}^{\mathcal{M}}$ in case $n = 0$.

In case $\mathcal{M}$ is squashed or $n > 0$ we can set

$$[\langle a, f\rangle]_E^{\mathcal{M}} \in \dot{F}^{\mathrm{Ult}(\mathcal{M},E)} \text{ iff } \{\,\bar{\alpha} \mid f(\bar{\alpha}) \in \dot{F}^{\mathcal{M}}\,\} \in E_a\,,$$

using amenability in the squashed $n = 0$ case. We are left with the case $\mathcal{M}$ is active and $n = 0$. Let $\mu = \mathrm{crit}\, \dot{F}^{\mathcal{M}}$. Let also $\eta = [\langle b, f\rangle]_E^{\mathcal{M}} \in \mathrm{OR} \cap \mathrm{Ult}_0(\mathcal{M}, E)$, and $h = [\langle b, g\rangle]_E^{\mathcal{M}}$, where h is a function with domain $i_E(\mu)$.

We want to put (a, h, η) into $\dot{F}^{\mathrm{Ult}(\mathcal{M},E)}$ for exactly one a. We may assume without loss of generality that $\mathrm{ran}\, h \subseteq \bigcup_n P([i_E(\mu)]^n)$.

Case 1. $\mu < \kappa$. So g is constant a.e.; in fact $g(\bar{x}) = h$ for almost every $\bar{x}$. Let $\gamma = \sup(\mathrm{ran}\, f \cap \mathrm{OR}^{\mathcal{M}})$, and let c be such that $\dot{F}^{\mathcal{M}}(c, h, \gamma)$. Using c we can compute k inside of $\mathcal{M}$:

$$k(\bar{x}) = \text{the unique } d \text{ such that } \dot{F}^{\mathcal{M}}(d, h, f(\bar{x}))\,.$$

Thus $k \in |\mathcal{M}|$. We then put

$$([\langle b, k\rangle]^{\mathcal{M}}_E, h, \eta) \in \dot{F}^{\mathrm{Ult}_0(\mathcal{M},E)}.$$

Case 2. $\kappa \leq \mu$. Let ℓ be a function in $|\mathcal{M}|$ with domain $= \mu$ and

$$\operatorname{ran} \ell = \bigcup_{\bar{x} \in [\kappa]^{\operatorname{card} b}} (\operatorname{ran} g(\bar{x}) \cap \bigcup_{n<\omega} P([\mu]^n)).$$

(We may assume $\operatorname{dom} g(\bar{x}) = \mu$ all $\bar{x}$.) Let $\gamma = \sup(\operatorname{ran} f \cap \mathrm{OR}^{\mathcal{M}})$. Let c be such that $\dot{F}^{\mathcal{M}}(c, \ell, \gamma)$. Using c we can compute in $\mathcal{M}$

$$k(\bar{x}) = \text{the unique } d \text{ such that } \dot{F}^{\mathcal{M}}(d, g(\bar{x}), f(\bar{x})).$$

So $k \in |\mathcal{M}|$. We then put

$$([\langle b, k\rangle]^{\mathcal{M}}_E, h, \eta) \in \dot{F}^{\mathrm{Ult}_0(\mathcal{M},E)}.$$

This completes the definition of $\mathrm{Ult}_n(\mathcal{M}, E)$. Notice the definition guarantees Los' Theorem holds for atomic formulae of $\mathcal{L} - \{\dot{v}, \dot{\gamma}\}$ (resp. $\mathcal{L}^*$).

Theorem 4.1. (*Los' Theorem*). *Let $n \geq 0$, let $\mathcal{M}$ be a ppm or sppm, and let E be a (κ, λ) pre-extender over $\mathcal{M}$, where $\kappa < \rho^{\mathcal{M}}_n$. Let $[a_i, f_i]^{\mathcal{M}}_E$ be an element of* $\mathrm{Ult}_n(\mathcal{M}, E)$ *for each $i \leq k$, and let $b = \bigcup_{i \leq k} a_i$. Then*

$$\mathrm{Ult}_n(\mathcal{M}, E) \models \varphi[[a_0, f_0]^{\mathcal{M}}_E, \ldots, [a_k, f_k]^{\mathcal{M}}_E]$$
$$\text{iff } \exists B \in E_b \forall \bar{u} \in B\, \mathcal{M} \models \varphi[\tilde{f}_0(\bar{u}), \ldots, \tilde{f}_k(\bar{u})]$$

for any generalized $r\Sigma_n$ (resp. $q\Sigma_n$) *formula φ. Here $\tilde{f}_i$ comes from f_i by adding the appropriate dummy variables.*

Remark. Assume $\mathcal{M}$, etc., are as in the hypotheses. If $n > 0$, then

$$A = \{\bar{u} \in [\kappa]^{\operatorname{card} b} \mid \mathcal{M} \models \varphi[\tilde{f}_0(\bar{u}) \cdots \tilde{f}_k(\bar{u})]\}$$

is in $\mathcal{M}$ as $\kappa < \rho^{\mathcal{M}}_n$.

If $n = 0$, then $A \notin |\mathcal{M}|$ is possible. However, our proof will show there is a $B \in E_b$ (so $B \in |\mathcal{M}|$) such that $B \subseteq A$ or $B \cap A = \varnothing$.

Proof. We consider only the case that $\mathcal{M}$ is a ppm (passive or active type I or II) as sppm behave exactly like passive ppm here.

Suppose first that φ is $r\Sigma_0$. If $n > 0$ we get the desired conclusion easily as there are enough functions defined by terms in Sk_n. So suppose $n = 0$.

For any $r\Sigma_0$ formula $\varphi = \varphi(v_0 \cdots v_k)$ and functions $f_0 \cdots f_k \in |\mathcal{M}|$ such that $\operatorname{dom} f_i = [\kappa]^{\operatorname{card} b}$ for all $i \leq k$ (where $b \subset \lambda$ is finite), we let

$$\bar{u} \in A_{\varphi,\bar{f}} \text{ iff } \mathcal{M} \models \varphi[f_0(\bar{u}) \cdots f_k(\bar{u})].$$

We show by induction on φ that there is a set $B \in E_b$ (so $B \in |\mathcal{M}|$) such that

$$B \subseteq A_{\varphi,\bar{f}} \text{ or } B \cap A_{\varphi,\bar{f}} = \varnothing$$

and

$$B \subseteq A_{\varphi,\bar{f}} \quad \text{iff} \quad \operatorname{Ult}_0(\mathcal{M}, E) \models \varphi[[\langle b, f_0 \rangle]_E^{\mathcal{M}} \cdots [\langle b, f_k \rangle]_E^{\mathcal{M}}].$$

For formulas φ which are Σ_0 in $\mathcal{L}$, a subinduction on Σ_0 in $\mathcal{L}$ formulas (using amenability) gives the result as usual. For $\varphi = \dot{F}(v_0, v_1, v_2)$, the construction of $\dot{F}^{\operatorname{Ult}_0(M,E)}$ guarantees the desired result. If φ is built from simpler $r\Sigma_0$ formulae by $\wedge$, $\vee$, or $\neg$ the inductive step is easy. Suppose $\varphi(v_0 \cdots v_k) =$ "v_0 is finite $\wedge$ $(\exists v_{k+1} \in v_0)\, \theta(v_0 \cdots v_{k+1})$". We may assume $f_0(\bar{u})$ is finite E_b a.e. as otherwise $B = \{\bar{u} \mid f_0(\bar{u}) \text{ infinite}\}$ does the job. But then we can fix $\ell \in \omega$ such that $\operatorname{card} f_0(\bar{u}) = \ell$ for E_b a.e. $\bar{u}$, and functions $g_1 \cdots g_\ell$ with $\operatorname{dom} = [\kappa]^{\operatorname{card} b}$ such that $f_0(\bar{u}) = \{g_1(\bar{u}) \cdots g_\ell(\bar{u})\}$ for E_b a.e. $\bar{u}$, say for $\bar{u} \in C$ where $C \in E_b$. Let B_i satisfy the induction hypothesis for $A_{\theta,\bar{f}^\frown g_i}$, and let $B = C \cap \bigcap_{i \leq \ell} B_i$. Then B works for $A_{\varphi,\bar{f}}$.

This completes the proof of 4.1 in the case that φ is $r\Sigma_0$.

We now show by induction on $i \leq n$ that 4.1 holds when φ is $r\Sigma_i$. We have done the case $i = 0$. The case φ is $r\Sigma_1$ and does not involve $\dot{\mu}$, $\dot{\nu}$, or $\dot{\gamma}$ now follows by the usual argument as there are enough functions defined by terms in Sk_n. But then 2.6 (b) implies that $\mathcal{P} = \operatorname{Ult}_n(\mathcal{M}, E)$ is a ppm of the same type as $\mathcal{M}$, and that $i_E(\mu^{\mathcal{M}}) = \mu^{\mathcal{P}}$, $i_E(\nu^{\mathcal{M}}) = \nu^{\mathcal{P}}$, $i_E(\gamma^{\mathcal{M}}) = \gamma^{\mathcal{P}}$. This gives 4.1 for arbitrary $r\Sigma_1$ formulae φ.

So now let $i > 1$. Notice first that as the relation $\operatorname{Th}_{i-1}^{Q}(a) = b$ is $\prod_1$ over $r\Sigma_{i-1}$ definable over Q, uniformly over all ppm Q, and as we have 4.1 for $\prod_1$ over $r\Sigma_{i-1}$ formulae by induction hypothesis and the fact that there are enough functions given by terms in Sk_n, we have (for $\operatorname{Ult} = \operatorname{Ult}_n(\mathcal{M}, E)$.),

(*) $\qquad \operatorname{Th}_{i-1}^{\operatorname{Ult}}([a, f]) = [b, g]$ iff for $E_{a \cup b}$ a.e. $\bar{x}$, $\operatorname{Th}_{i-1}^{\mathcal{M}}(\tilde{f}(\bar{x})) = \tilde{g}(\bar{x})$.

Let $\pi : \mathcal{M} \to \operatorname{Ult}_n(\mathcal{M}, E)$ be the canonical embedding. It follows that

(**)
$$\rho_{i-1}^{\operatorname{Ult}} = \begin{cases} \mathrm{OR}^{\mathcal{M}} & \text{if } \rho_{i-1}^{\mathcal{M}} = \mathrm{OR}^{\operatorname{Ult}} \\ \pi(\rho_{i-1}^{\mathcal{M}}) & \text{otherwise.} \end{cases}$$

We prove the case $\rho_{i-1}^{\operatorname{Ult}} = \pi(\rho_{i-1}^{\mathcal{M}})$. Suppose $\rho_{i-1}^{\mathcal{M}} < \mathrm{OR}^{\mathcal{M}}$. We show first that $\pi(\rho_{i-1}^{\mathcal{M}}) \leq \rho_{i-1}^{\operatorname{Ult}}$. For let $\alpha = [a, f]_E^{\mathcal{M}} < \pi(\rho_{i-1}^{\mathcal{M}})$, and let $q = [a, g]_E^{\mathcal{M}}$. We may assume $f(\bar{x}) < \rho_{i-1}^{\mathcal{M}}$ for all $\bar{x}$. Define

$$\begin{aligned} h(\bar{x}) &= \operatorname{Th}_{i-1}^{\mathcal{M}}(f(\bar{x}) \cup \{g(\bar{x})\}) \\ &= \text{least } b \text{ such that } \dot{T}_{i=1}^{\mathcal{M}}(\langle f(\bar{x}), g(\bar{x}) \rangle, b). \end{aligned}$$

Then h is one of the functions used in forming $\mathrm{Ult}_n(\mathcal{M}, E)$, and as we observed above in (*)

$$\mathrm{Th}^{\mathrm{Ult}}_{i-1}(\alpha \cup \{q\}) = [a, h]^{\mathcal{M}}_E .$$

Thus $\alpha < \rho^{\mathrm{Ult}}_{i-1}$. Thus $\pi(\rho^{\mathcal{M}}_{i-1}) \leq \rho^{\mathrm{Ult}}_{i-1}$.

On the other hand, pick $q \in |\mathcal{M}|$ such that $\mathrm{Th}^{\mathcal{M}}_{i-1}(\rho^{\mathcal{M}}_{i-1} \cup \{q\}) \notin |\mathcal{M}|$. Then by (*) $\mathrm{Th}^{\mathrm{Ult}}_{i-1}(\pi(\rho^{\mathcal{M}}_{i-1}) \cup \{\pi(q)\}) \notin \mathrm{Ult}$. Thus $\rho^{\mathrm{Ult}}_{i-1} \leq \pi(\rho^{\mathcal{M}}_{i-1})$.

Putting (*) and (**) together, we have

$$\dot{T}^{\mathrm{Ult}}_{i-1}([a, f], [b, g]) \text{ iff for } E_{a\cup b} \quad \text{a.e. } \bar{x},\ \dot{T}^{\mathcal{M}}_{i-1}(\tilde{f}(\bar{x}), \tilde{g}(\bar{x})) .$$

Now suppose

$$\varphi(\bar{v}) = \exists a \exists b (\dot{T}_{i-1}(a, b) \wedge \psi(a, b, \bar{v}))$$

where ψ is $r\Sigma_1$. We check one direction of the conclusion of 4.1. Suppose that for E_b a.e. $\bar{u}$, $\mathcal{M} \models \varphi[f_0(\bar{u}) \cdots f_k(\bar{u})]$. Let $f_i = f_{\tau_i q_i}$ where $\tau_i \in \mathrm{Sk}_n$ and $q_i \in |\mathcal{M}|$. We can translate "$\psi(a, b, \bar{v}) \wedge \dot{T}_{i-1}(a, b)$" into an $r\Sigma_n$ formula; this gives us terms σ_0 and σ_1 in Sk_n which Skolemize the result, i.e., such that for E_b a.e. $\bar{u}$

$$\mathcal{M} \models \psi(\sigma_0(\bar{u}, \bar{q}), \sigma_1(\bar{u}, \bar{q}), \tau_0(\bar{u}, q_0) \cdots \tau_k(\bar{u}, q_k)) \wedge \dot{T}_{i-1}(\sigma_0(\bar{u}, \bar{q}), \sigma_1(\bar{u}, \bar{q})) ,$$

where $\bar{q} = \langle q_1 \cdots q_k \rangle$. But then, letting $g_0 = f_{\sigma_0, \bar{q}}$ and $g_1 = f_{\sigma_1, \bar{q}}$,

$$\mathrm{Ult}_n(\mathcal{M}, E) \models \dot{T}_{i-1}([b, g_0], [b, g_1]) \wedge \psi([b, g_0], [b, g_1][b, f_0] \cdots [b, f_k])$$

as desired.

Finally, we prove 4.1 in the case φ is generalized $r\Sigma_n$ with $n > 0$. Notice first that if $\tau(v_0 \cdots v_k) \in \mathrm{Sk}_n$, then

$$\tau^{\mathrm{Ult}}[[a, f_0], \ldots, [a, f_k]] = [a, \lambda \bar{u} \cdot \tau^{\mathcal{M}}[f_0(\bar{u}) \cdots f_k(\bar{u})]]^{\mathcal{M}}_E$$

for any $[a, f_0] \cdots [a, f_k] \in \mathrm{Ult} = \mathrm{Ult}_n(\mathcal{M}, E)$. To see this, it is enough to consider the basic terms $\tau_\theta \in \mathrm{Sk}_n$. But the graph of such a term is definable by a Boolean combination of $r\Sigma_n$ formulae, uniformly over all ppm, so we can use the term-free case of 4.1 just proved.

But now if $\varphi(v)$ is $r\Sigma_n$ and $\tau(v) \in \mathrm{Sk}_n$ then

$$\begin{aligned} \mathrm{Ult} \models \varphi(\tau(v))[[a, f]^{\mathcal{M}}_E] &\text{ iff } \mathrm{Ult} \models \varphi[\tau^{\mathrm{Ult}}[[a, f]^{\mathcal{M}}_E]] \\ &\text{ iff } \mathrm{Ult} \models \varphi[[a, \lambda u \cdot \tau^{\mathcal{M}}[f(\bar{u})]]^{\mathcal{M}}_E] \\ &\text{ iff for } E_a \text{ a.e. } \bar{u}, \mathcal{M} \models \varphi[\tau^m[f(\bar{u})]] \\ &\text{ iff for } E_a \text{ a.e. } \bar{u}, \mathcal{M} \models \varphi(\tau(v))[f(\bar{u})] \end{aligned}$$

as desired. Of course, the case φ or τ having more variables involves only more notation. □

In the course of proving 4.1 we have shown

Corollary 4.2. *Let $\mathcal{M}$, etc., be as in the hypotheses of* 4.1, *and let* $\pi : \mathcal{M} \to \mathrm{Ult}_n(\mathcal{M}, E)$ *be the canonical embedding. Then for* $i < n$

$$\rho_i^{\mathcal{M}} < \mathrm{OR}^{\mathcal{M}} \quad \text{iff } \rho_i^{\mathrm{Ult}} < \mathrm{OR}^{\mathrm{Ult}}$$

and

$$\rho_i^{\mathcal{M}} < \mathrm{OR}^{\mathcal{M}} \Rightarrow \pi(\rho_i^{\mathcal{M}}) = \rho_i^{\mathrm{Ult}}$$

(*where* $\mathrm{Ult} = \mathrm{Ult}_n(\mathcal{M}, E)$).

We would like to show that under the hypotheses of 4.1, the canonical $\pi : \mathcal{M} \to \mathrm{Ult}_n(\mathcal{M}, E)$ is generalized $r\Sigma_{n+1}$ elementary. For this we seem to need (essentially) that $\mathcal{M}$ be n-sound. Fortunately, we shall never want to form $\mathrm{Ult}_n(\mathcal{M}, E)$ unless $\mathcal{M}$ is n-sound.

Corollary 4.3. *Let $\mathcal{M}$, etc., be as in the hypotheses of* 4.1, *and let* $\pi : \mathcal{M} \to \mathrm{Ult}_n(\mathcal{M}, E)$ *be the canonical embedding. Suppose that for some* $p \in |\mathcal{M}|$, $\mathcal{M} = \mathcal{H}_n^{\mathcal{M}}(\rho_n^{\mathcal{M}} \cup \{p\})$. *Then π is generalized $r\Sigma_{n+1}$ (resp. $q\Sigma_{n+1}$) elementary; moreover* $\rho_n^{\mathrm{Ult}(\mathcal{M},E)} = \sup \pi'' \rho_n^{\mathcal{M}}$.

PROOF. Let $\mathrm{Ult} = \mathrm{Ult}_n(\mathcal{M}, E)$. We show first that $\sup \pi'' \rho_n^{\mathcal{M}} \geq \rho_n^{\mathrm{Ult}}$; for this it is enough to show that if $\mathcal{H}_n^{\mathcal{M}}(\rho_n^{\mathcal{M}} \cup \{p\}) = \mathcal{M}$, then

$$\mathcal{H}_n^{\mathrm{Ult}}(\sup \pi'' \rho_n^{\mathcal{M}} \cup \{\pi(p)\}) = \mathrm{Ult}.$$

(For then $\mathrm{Th}_n^{\mathrm{Ult}}(\sup \pi'' \rho_n^{\mathcal{M}} \cup \{\pi(p)\}) \notin \mathrm{Ult}$ by a diagonal argument.) So let $\mathcal{H}_n^{\mathcal{M}}(\rho_n^{\mathcal{M}} \cup \{p\}) = \mathcal{M}$, and let $[a, f] \in \mathrm{Ult}$. Then there is a term $\tau \in \mathrm{Sk}_n$ (resp. SK_n) and parameters $\bar{b} \in [\rho_n^{\mathcal{M}} \cup \{p\}]^{<\omega}$ such that for all $\bar{u}$, $f(\bar{u}) = \tau^{\mathcal{M}}[\bar{u}, \bar{b}]$. Let $\mathrm{id}(\bar{u}) = \bar{u}$, $c_{\bar{b}}(\bar{u}) = \bar{b}$. By the Los Theorem, $[a, f]_E^{\mathcal{M}} = \tau^{\mathrm{Ult}}[[a, \mathrm{id}]_E^{\mathcal{M}}, [a, c_{\bar{b}}]_E^{\mathcal{M}}] = \tau^{\mathrm{Ult}}[a, \pi(\bar{b})]$. Since $a \in [\pi(\kappa)]^{<\omega}$ and $\kappa < \rho_n^{\mathcal{M}}$, and since $\pi(\bar{b}) \in [\sup \pi'' \rho_n^{\mathcal{M}} \cup \{\pi(p)\}]^{<\omega}$, we're done.

We claim next that $\rho_n^{\mathrm{Ult}} \geq \sup \pi'' \rho_n^{\mathcal{M}}$. For by the Los Theorem we have easily that for $a, b \in |\mathcal{M}|$

$$\mathrm{Th}_n^{\mathcal{M}}(a) = b \quad \text{iff} \quad \mathrm{Th}_n^{\mathrm{Ult}}(\pi(a)) = \pi(b)$$

[For the "only if" direction, let $\bar{c} \in \pi(a)^{<\omega}$ and φ be generalized $r\Sigma_n$. Let $\bar{c} = [d, f]_E^{\mathcal{M}}$. Then $(\varphi, \bar{c}) \in \mathrm{Th}_n^{\mathrm{Ult}}(\pi(a))$ iff (for E_d a.e. $\bar{u}$ $(\varphi, f(\bar{u})) \in \mathrm{Th}_n^{\mathcal{M}}(a)$) iff $(\varphi, \bar{c}) \in \pi(b)$.]

It follows that $\forall \gamma < \sup \pi'' \rho_n^{\mathcal{M}}$, $\forall p \in |\mathcal{M}|$, $\mathrm{Th}_n^{\mathrm{Ult}}(\gamma \cup \{\pi(p)\}) \in \mathrm{Ult}$. Now fix p such that $\mathcal{H}_n^{\mathcal{M}}(\rho_n^{\mathcal{M}} \cup \{p\}) = \mathcal{M}$. Let $\alpha < \sup \pi'' \rho_n^{\mathcal{M}}$ and $r \in \mathrm{Ult}$; we must see that $\mathrm{Th}_n^{\mathrm{Ult}}(\alpha \cup \{r\}) \in \mathrm{Ult}$. Fix $\gamma < \sup \pi'' \rho_n^{\mathcal{M}}$ such that $\alpha < \gamma$ and for some $\bar{\beta} \in \gamma^{<\omega}$, and some $\tau \in \mathrm{Sk}_n$, $r = \tau^{\mathrm{Ult}}[\bar{\beta}, \pi(p)]$. Then $\mathrm{Th}_n^{\mathrm{Ult}}(\gamma \cup \{\pi(p)\})$ is in Ult, and from it we can compute $\mathrm{Th}_n^{\mathrm{Ult}}(\alpha \cup \{r\})$ inside Ult.

Thus $\rho_n^{\mathrm{Ult}} = \sup \pi'' \rho_n^{\mathcal{M}}$.

Now let $[a, f]_E^{\mathcal{M}} = \langle \alpha, q \rangle$, where $\alpha < \sup \pi'' \rho_n^{\mathcal{M}}$. We claim that for any g

$$\dot{T}_n^{\mathrm{Ult}}([a, f], [a, g]) \quad \text{iff for } E_a \text{ a.e. } \bar{u}\,,\ \dot{T}_n^{\mathcal{M}}(f(\bar{u}),\ g(\bar{u}))\,.$$

$\Leftarrow$ is easy since $\mathrm{Th}_n^{\mathcal{P}}(a) = b$ is uniformly Π_1 over $r\Sigma_n^{\mathcal{P}}$, and we have the Los Theorem for $r\Sigma_n$ formulae. So suppose $\dot{T}_n^{\mathrm{Ult}}([a, f], [a, g])$. Let $q = \tau^{\mathrm{Ult}}[\bar{\beta}, \pi(p)]$ where $\bar{\beta} \in [\sup \pi'' \rho_n^{\mathcal{M}}]^{<\omega}$ and pick $\gamma < \rho_n^{\mathcal{M}}$ such that $\bar{\beta}, \alpha < \pi(\gamma)$. Let $b = \mathrm{Th}_n^{\mathcal{M}}(\gamma \cup \{p\})$, so that

$$\pi(b) = \mathrm{Th}_n^{\mathrm{Ult}}(\pi(\gamma) \cup \{\pi(p)\})\,.$$

Then we have

(1) $(\varphi, \bar{c}) \in [a, g]$ iff

$$\varphi \text{ is generalized } r\Sigma_n \text{ and } \bar{c} \in [\alpha \cup \{q\}]^{<\omega} \text{ and } (\varphi^*, \bar{c}^*) \in \pi(b),$$

where $\varphi^*(\bar{c}^*)$ is the obvious way of rewriting $\varphi(\bar{c})$ so that the parameters $\bar{c}^*$ come from $\pi(\gamma) \cup \{\pi(p)\}$. Thus the map $(\varphi, \bar{c}) \mapsto (\varphi^*, \bar{c}^*)$ is $r\Delta_1$ over Ult in the parameters α, q, and $\bar{\beta}$. Let $\bar{\beta} = [a, h]_E^{\mathcal{M}}$, where we assume for notational convenience that the support is a (otherwise enlarge all supports). Then the fact that (1) holds in Ult is a $r\Pi_1$ fact about $[a, f]$, $[a, g]$, and $[a, h]$. It follows that for E_a a.e. $\bar{u}$,

$(\varphi, \bar{c}) \in g(\bar{u})$ iff

$$\varphi \text{ is generalized } r\Sigma_n \text{ and } \bar{c} \in [\alpha_{\bar{u}} \cup \{q_{\bar{u}}\}]^{<\omega} \text{ and } (\varphi^*, \bar{c}^*) \in b\,,$$

where $f(\bar{u}) = \langle \alpha_{\bar{u}}, q_{\bar{u}} \rangle$ and $(\varphi^*, \bar{c}^*)$ comes from rewriting $(\varphi, \bar{c})$ by substituting $\tau(h(\bar{u}), p)$ for occurrences of $q_{\bar{u}}$. But now

$$q_{\bar{u}} = \tau^{\mathcal{M}}[h(\bar{u}),\ p]$$

for E_a a.e. $\bar{u}$, by the Los Theorem. As $b = \mathrm{Th}_n^{\mathcal{M}}(\gamma \cup \{p\})$, we see

$$g(\bar{u}) = \mathrm{Th}_n^{\mathcal{M}}(\alpha_{\bar{u}} \cup \{q_{\bar{u}}\})$$

for E_a a.e. $\bar{u}$. As $\alpha_{\bar{u}} < \rho_n^{\mathcal{M}}$ a.e., we get

$$T_n^{\mathcal{M}}(f(\bar{u}),\ g(\bar{u}))$$

for E_a a.e. $\bar{u}$, as desired.

Finally, let

$$\varphi(\bar{v}) = \exists a \exists b\, (\dot{T}_n(a, b) \wedge \psi(a, b, \bar{v}))$$

be an $r\Sigma_{n+1}$ formulae. If $\mathcal{M} \models \varphi[\bar{x}]$, then we have a, b such that $\dot{T}_n^{\mathcal{M}}(a,b) \wedge \psi^{\mathcal{M}}(a,b,\bar{x})$, so $\dot{T}_n^{\mathrm{Ult}}(\pi(a),\pi(b))$ and $\psi^{\mathrm{Ult}}(\pi(a),\pi(b),\bar{x})$, so Ult $\models \varphi[\pi(\bar{x})]$. On the other hand if Ult $\models \varphi[\pi(\bar{x})]$, then we have a, f, g such that $[a,f] = \langle \alpha, q\rangle$ for some $\alpha < \sup \pi'' \rho_n^{\mathcal{M}}$, and

$$\mathrm{Ult} \models \dot{T}_n([a,f],[a,g]) \wedge \psi([a,f],[a,g],\pi(\bar{x}))\,.$$

By our claim, for E_a a.e. $\bar{u}$

$$\mathcal{M} \models \dot{T}_n(f(\bar{u}), g(\bar{u})) \wedge \psi(f(\bar{u}), g(\bar{u}), \bar{x})\,.$$

Thus $\mathcal{M} \models \varphi[\bar{x}]$, as desired.

We can now show $\pi(\tau^{\mathcal{M}}(x)) = \tau^{\mathrm{Ult}}(\pi(x))$ for all $\tau \in \mathrm{Sk}_{n+1}$, since the graphs of basic terms in Sk_{n+1} are definable by boolean combinations of $r\Sigma_{n+1}$ formulae. It follows that π is generalized $r\Sigma_{n+1}$ elementary.

Relations to Dodd-Jensen.

It is easy to see that if $\mathcal{M}$ is n-sound, $\mathrm{Ult}_n(\mathcal{M},E)$ is exactly what is obtained by the Dodd-Jensen procedure of coding $\mathcal{M}$ onto $\rho_n^{\mathcal{M}}$, taking a Σ_0 ultrapower of the coded structure, and then decoding.

For let $\mathcal{M}$ be a ppm or sppm, $n \geq 1$, and $\mathcal{M} = \mathcal{H}_n^{\mathcal{M}}(\rho_n^{\mathcal{M}} \cup \{q\})$. Let

$$\pi : \mathcal{M} \to \mathrm{Ult}_n(\mathcal{M},E) = \mathcal{N}$$

be the canonical embedding. Now let

$$A^{\mathcal{M}} = \mathrm{Th}_n^{\mathcal{M}}(\rho_n^{\mathcal{M}} \cup \{q\}),\ \text{coded as a subset of } \rho_n^{\mathcal{M}}\,,$$
$$A^{\mathcal{N}} = \mathrm{Th}_n^{\mathcal{N}}(\rho_n^{\mathcal{N}} \cup \{\pi(q)\}),\ \text{similarly coded}\,.$$

Let

$$\mathcal{P} = (J_{\rho_n^{\mathcal{M}}}^{\dot{E}^{\mathcal{M}}}, \in, \dot{E}^{\mathcal{M}} \restriction \rho_n^{\mathcal{M}}, A^{\mathcal{M}})$$
$$\mathcal{Q} = (J_{\rho_n^{\mathcal{N}}}^{\dot{E}^{\mathcal{N}}}, \in, \dot{E}^{\mathcal{N}} \restriction \rho_n^{\mathcal{N}}, A^{\mathcal{N}})$$

be the master code structures associated to $\mathcal{M}$ and $\mathcal{N}$. Then

$$\pi : \mathcal{P} \xrightarrow{\Sigma_0} \mathcal{Q}$$

cofinally; this is contained in 4.3. Note also that if $[a,f]_E^{\mathcal{M}} \in |\mathcal{Q}|$, then $\exists_\beta < \rho_n^{\mathcal{M}}$ such that $f(\bar{u}) < \beta$ E_a a.e., so since f is given by a term in Sk_n, in fact $f \in |\mathcal{M}|$ and hence $f \in |\mathcal{P}|$. So in fact

$$\mathcal{Q} = \mathrm{Ult}_0(\mathcal{P},E)$$

and $\pi \restriction |\mathcal{P}|$ is the canonical embedding for this Σ_0 ultrapower. Notice finally that all of $\mathcal{N}$ can be decoded from $\mathcal{Q}$, since $\mathcal{N} = \mathcal{H}_n^{\mathcal{N}}(\rho_n^{\mathcal{N}} \cup \{\pi(q)\})$.

Although we can make sense of $\mathrm{Ult}_n(\mathcal{M},E)$ in the case $\mathcal{M}$ is not n-sound, in practice we shall never need to form such an ultrapower. Thus our construction of $\mathrm{Ult}_n(\mathcal{M},E)$ does not go beyond Dodd-Jensen in any important way.

We describe now the preservation of the core parameters $p_i(\mathcal{M})$, for $i \leq n$, in the case that $\mathcal{M}$ is n-sound.

Lemma 4.4. *Let $\mathcal{M}$ be n-sound, let E be an extender over $\mathcal{M}$ with* $\mathrm{crit}(E) < \rho_n^{\mathcal{M}}$, *and let* $\pi : \mathcal{M} \to \mathrm{Ult}_n(\mathcal{M}, E)$ *be the canonical embedding. Then*

(a) $\mathrm{Ult}_n(\mathcal{M}, E)$ *is n-sound, and*

(b) π *is an n-embedding.*

PROOF. Let $\mathcal{N} = \mathrm{Ult}_n(\mathcal{M}, E)$. It is enough to show that for all $i \leq n$

$$\mathfrak{C}_i(\mathcal{N}) = \mathcal{N},$$

and

$$p_i(\mathcal{N}) = \pi(p_i(\mathcal{M})).$$

For then by soundness $\rho_i(\mathcal{N}) = \rho_i^{\mathcal{N}}$ for all $i \leq n$, and similarly for $\mathcal{M}$, so that π maps the core projecta properly by 4.2 and 4.3.

We proceed by induction on $i \leq n$. For $i = 0$ there is nothing to prove. Now let $i = 1$ and Let r be the first standard parameter of $\mathcal{M}$. Thus as $\mathcal{M}$ is 1-sound, $p_1(\mathcal{M}) = \langle r, \varnothing \rangle$ and r is 1-solid and 1-universal over $\mathcal{M}$.

Let $r = \langle \alpha_0 \cdots \alpha_\ell \rangle$, and

$$b_j = \mathrm{Th}_1^{\mathcal{M}}(\alpha_j \cup \{\alpha_0 \cdots \alpha_{j-1}\}), \quad 0 \leq j \leq \ell$$

so that $b_j \in |\mathcal{M}|$ by solidity. By 4.3, π is at least $r\Sigma_2$ elementary, so

$$\pi(b_j) = \mathrm{Th}_1^{\mathcal{N}}(\pi(\alpha_j) \cup \{\pi(\alpha_0) \cdots \pi(\alpha_{j-1})\}).$$

It follows that no $s <_{\mathrm{lex}} \pi(r)$ can serve as the 1st standard parameter of $\mathcal{N}$. On the other hand, $\mathrm{Th}_1^{\mathcal{N}}(\rho_1^{\mathcal{N}} \cup \{\pi(r)\}) \notin |\mathcal{N}|$ and in fact $\mathcal{H}_1^{\mathcal{N}}(\rho_1^{\mathcal{N}} \cup \{\pi(r)\}) = |\mathcal{N}|$. [If $n = 1$ this is implicit in the proof of 4.3. Suppose $n > 1$. Let $[a, f]$ by an arbitrary element of $|\mathcal{N}|$. Notice that if we let, for $x \in |\mathcal{M}|$,

$$h(x) = 1\text{ st} \quad \text{(in order of construction) } (\tau, \bar{\beta}) \text{ such that}$$
$$\tau \in \mathrm{Sk}_1 \ \wedge \bar{\beta} \in (\rho_1^{\mathcal{M}})^{<\omega} \wedge \tau^{\mathcal{M}}[\bar{\beta}, r] = x$$

then $h(x) = \sigma^{\mathcal{M}}[x, r]$ for some term $\sigma \in \mathrm{Sk}_2$. So if we let

$$g(\bar{u}) = h(f(\bar{u})),$$

then g is one of the functions used to form $\mathcal{N}$, and if $[a, g] = (\tau, \bar{\beta})$, then $\tau \in \mathrm{Sk}_1$ and $\bar{\beta} \in (\rho_1^{\mathcal{N}})^{<\omega}$ and $\tau^{\mathcal{N}}[\bar{\beta}, \pi(r)] = [a, f]$, as desired.]

So $\pi(r)$ is the 1st standard parameter of $\mathcal{N}$, is 1-solid and 1-universal and $\mathcal{N}$ is 1-sound. As $\mathcal{N}$ is 1-sound, $p_1(\mathcal{N}) = \langle \pi(r), \varnothing \rangle = \pi(p_1(\mathcal{M}))$, as desired.

The case $i > 1$ of the induction involves a bit more notation but no new ideas, so we omit it. □

The next question, clearly, is how π moves $\rho_{n+1}^{\mathcal{M}}$ and the $n+1$st standard parameter of $(\mathcal{M}, p_n(\mathcal{M}))$. We answer this under a solidity hypothesis. Our proof is in essence drawn from Mitchell [M?]; it was recently re-discovered by S. Friedman.

We must also, it seems, impose an additional condition on E.

DEFINITION 4.4.1. Let $\mathcal{M}$ be a ppm or an sppm, and E a (κ, λ) extender over $\mathcal{M}$. Then E is *close to* $\mathcal{M}$ iff for every $a \in [\lambda]^{<\omega}$

(1) E_a is $r\Sigma_1^{\mathcal{M}}$ (resp. $q\Sigma_1^{\mathcal{M}}$) and
(2) if $A \in |\mathcal{M}|$ and $\mathcal{M} \models \text{card } A \leq \kappa$, then $E_a \cap A \in |\mathcal{M}|$.

The purpose of this restriction on E is explained by the following lemma.

Lemma 4.5. *Suppose E is a (κ, λ) extender which is close to $\mathcal{M}$, and $\kappa < \rho_n^{\mathcal{M}}$. Then*

$$P(\kappa) \cap |\mathcal{M}| = P(\kappa) \cap |\text{Ult}_n(\mathcal{M}, E)| .$$

If, in addition, $\mathcal{M} = \mathcal{H}_n^{\mathcal{M}}(\rho_n^{\mathcal{M}} \cup \{q\})$ for some q and $\rho_{n+1}^{\mathcal{M}} \leq \kappa$, then

$$\rho_{n+1}^{\mathcal{M}} = \rho_{n+1}^{\text{Ult}_n(\mathcal{M},E)} .$$

PROOF. The nontrivial part of the first sentence is the assertion that $P(\kappa) \cap |\text{Ult}_n(\mathcal{M}, E)| \subseteq |\mathcal{M}|$. So let $[a, f]_E^{\mathcal{M}} \subseteq \kappa$, where $f : [\kappa]^i \to P(\kappa)$. Since $\kappa < \rho_n^{\mathcal{M}}$ and f is defined from a parameter by a term in Sk_n, in fact $f \in |\mathcal{M}|$. For $\alpha < \kappa$, let $A_\alpha = \{\bar{u} \in [\kappa]^i \mid \alpha \in f(\bar{u})\}$. From $E_a \cap \{A_\alpha \mid \alpha < \kappa\}$ we can compute $[a, f]_E^{\mathcal{M}}$. Since E is close to $\mathcal{M}$, we get $[a, f]_E^{\mathcal{M}} \in |\mathcal{M}|$.

For the second assertion it is convenient to use the master code structures. Let $\mathcal{N} = \text{Ult}_n(\mathcal{M}, E)$ and fix q such that $\mathcal{M} = \mathcal{H}_n^{\mathcal{M}}(\rho_n^{\mathcal{M}} \cup \{q\})$. Set

$$\begin{aligned} A^{\mathcal{M}} &= \text{Th}_n^{\mathcal{M}}(\rho_n^{\mathcal{M}} \cup \{q\}) \text{, coded as a subset of } \rho_n^{\mathcal{M}} , \\ A^{\mathcal{N}} &= \text{Th}_n^{\mathcal{N}}(\rho_n^{\mathcal{N}}(\rho_n^{\mathcal{N}} \cup \{\pi(q)\}) \text{, coded as a subset of } \rho_n^{\mathcal{N}} , \\ \mathcal{P} &= (J_{\rho_n^{\mathcal{M}}}^{\dot{E}^{\mathcal{M}}}, \in, \dot{E}^{\mathcal{M}} \restriction \rho_n^{\mathcal{M}}, A^{\mathcal{M}}), \\ \mathcal{Q} &= (J_{\rho_n^{\mathcal{N}}}^{\dot{E}^{\mathcal{N}}}, \in, \dot{E}^{\mathcal{N}} \restriction \rho_n^{\mathcal{N}}, A^{\mathcal{N}}), \end{aligned}$$

so that $\mathcal{Q} = \text{Ult}_0(\mathcal{P}, E)$ with canonical embedding π, which is cofinal and Σ_1 elementary. If $n = 0$ then we take $\mathcal{P} = \mathcal{M}$ and $\mathcal{Q} = \mathcal{N}$.

By Lemmas 2.10 and 2.11, $\rho_{n+1}^{\mathcal{M}}$ is the least α such that some Σ_1 over $\mathcal{P}$ set $B \subseteq \alpha$ is such that $B \notin |\mathcal{P}|$, and similarly for $\rho_{n+1}^{\mathcal{N}}$ and $\mathcal{Q}$.

To see that $\rho_{n+1}^{\mathcal{N}} \leq \rho_{n+1}^{\mathcal{M}}$, let $B \subseteq \rho_{n+1}^{\mathcal{M}}$ be Σ_1 over $\mathcal{P}$ and $B \notin |\mathcal{P}|$. Since $\rho_{n+1}^{\mathcal{M}} \leq \kappa$ and π is Σ_1 elementary, B is Σ_1 over $\mathcal{Q}$. But $P(\kappa) = P(\kappa)^{\mathcal{M}} = P(\kappa)^{\mathcal{N}} = P(\kappa)^{\mathcal{Q}}$ using the first assertion of 4.5 and strong acceptability. Thus $B \notin |\mathcal{Q}|$.

To see that $\rho_{n+1}^{\mathcal{M}} \leq \rho_{n+1}^{\mathcal{N}}$, let $\alpha \leq \rho_{n+1}^{\mathcal{M}}$ and $B \subseteq \alpha$ be Σ_1 over $\mathcal{Q}$. It is enough to show that B is Σ_1 over $\mathcal{P}$. Let

$$\eta \in B \Leftrightarrow \mathcal{Q} \models \psi[\eta, [a, f]]$$

where ψ is Σ_1 in the language of $\mathcal{Q}$. For $\delta < \mathrm{OR}^{\mathcal{Q}}$ let

$$\mathcal{Q}_\delta = (J_\delta^{\dot{E}^{\mathcal{N}}}, \in, \dot{E}^{\mathcal{N}} \restriction \omega\delta, A^{\mathcal{N}} \cap \omega\delta)$$

and similarly define $\mathcal{P}_\delta$ for $\delta < \mathrm{OR}^{\mathcal{P}}$. So

$$\eta \in B \Leftrightarrow \exists \delta < \mathrm{OR}^{\mathcal{Q}} \quad \mathcal{Q}_\delta \models \psi[\eta, [a, f]]$$

so

$$\begin{aligned} \eta \in B &\Leftrightarrow \exists \delta < \mathrm{OR}^{\mathcal{P}}\ \mathcal{Q}_{\pi(\delta)} \models \psi[\eta, [a, f]] \\ &\Leftrightarrow \exists \delta < \mathrm{OR}^{\mathcal{P}} \exists X \in E_a\ \forall \bar{u} \in X \mathcal{P}_\delta \models \psi[\eta, f(\bar{u})] . \end{aligned}$$

Now as E is close to $\mathcal{M}$, the E_a is an $r\Sigma_1^{\mathcal{M}}$ subset of $|\mathcal{P}|$. By Lemma 2.11, E_a is Σ_1 over $\mathcal{P}$. Thus B is Σ_1 over $\mathcal{P}$, as desired. □

We now consider preservation of the $n+1$st standard parameter.

Lemma 4.6. *Let $\mathcal{M}$ be a ppm or sppm, $n \geq 0$, and $\mathcal{M} = \mathcal{H}_n^{\mathcal{M}}(\rho_n^{\mathcal{M}} \cup \{q\})$ if $n \geq 1$. Let E be an extender close to $\mathcal{M}$ such that $\rho_{n+1}^{\mathcal{M}} \leq \mathrm{crit}\ E < \rho_n^{\mathcal{M}}$. Let*

$$\pi : \mathcal{M} \to \mathrm{Ult}_n(\mathcal{M}, E) = \mathcal{N}$$

be the canonical embedding. Suppose that r is the $n+1$st standard parameter of $(\mathcal{M}, q)$ and that r is $n+1$-solid over $(\mathcal{M}, q)$.

Then $\pi(r)$ is the $n+1$st standard parameter of $(\mathcal{N}, \pi(q))$, and $\pi(r)$ is $n+1$-solid over $(\mathcal{N}, \pi(q))$.

Proof. We will give the proof for the case $n = 0$ with a passive premouse of limit length. The general proof is the same as this, using the fact that $r\Sigma_{n+1}$ is equivalent to Σ_1 over the appropriate master code structure. See lemma 2.11 for the case of $n > 0$ and the remark following corollary 2.2 for the case of $n = 0$ with an active premouse. For successor ordinals $\lambda = \gamma + 1$ write $\mathcal{M}_\lambda = \bigcup_{n \in \omega} S_{\omega\gamma+n}^{\mathcal{M}_\lambda}$, where $(S_\nu^{\mathcal{M}_\lambda} : \nu < \omega\lambda)$ is Jensen's S sequence, and use the same proof as below.

Let us consider first the case $n = 0$, $\mathcal{M}$ is passive, and

$$\mathcal{M} = (J_\lambda^{\dot{E}^{\mathcal{M}}}, \in, \dot{E}^{\mathcal{M}}) \quad (\lambda \text{ limit}) .$$

Now by 4.5, $\rho_1^{\mathcal{M}} = \rho_1^{\mathcal{N}}$ and

$$\mathrm{Th}_1^{\mathcal{N}}(\rho_1^{\mathcal{N}} \cup \{\pi(\sigma), \pi(q)\}) = \mathrm{Th}_1^{\mathcal{M}}(\rho_1^{\mathcal{M}} \cup \{r, q\}) \notin |\mathcal{N}| ,$$

so it is enough to show that $\pi(r)$ is 1-solid over $(\mathcal{N}, \pi(q))$. Let

$$\begin{aligned} r &= \langle \alpha_0 \cdots \alpha_\ell \rangle \\ b_i &= \mathrm{Th}_1^{\mathcal{M}}(\alpha_i \cup \{\alpha_0 \cdots \alpha_{i-1}, q\}) \cap \{(\varphi, \bar{c}) \mid \varphi \text{ is pure } r\Sigma_1\} \\ c_i &= \mathrm{Th}_1^{\mathcal{N}}(\pi(\alpha_i) \cup \{\pi(\alpha_0) \cdots \pi(\alpha_{i-1}),\ \pi(q)\}) \cap \{(\varphi, \bar{c}) \mid \varphi \text{ is pure } r\Sigma_1\} \end{aligned}$$

for $0 \le i \le \ell$. By 2.10, it will be enough to show $c_i \in |\mathcal{N}|$ for $0 \le i \le \ell$. So fix i such that $0 \le i \le \ell$.

For $\gamma < \lambda$ let

$$R_\gamma = b_i \cap \{(\varphi, \bar{c}) \mid \mathcal{M}_\gamma \models \varphi[\bar{c}]\}$$

so that $b_i = \bigcup_{\gamma<\lambda} R_\gamma$. Similarly, let

$$S_\gamma = c_i \cap \{(\varphi, \bar{c}) \mid \mathcal{N}_\gamma \models \varphi[\bar{c}]\}$$

so that $c_i = \bigcup_{\omega\gamma<\mathrm{OR}^{\mathcal{M}}} S_\gamma$. It is easy to see that $\pi(R_\gamma) = S_{\pi(\gamma)}$, and thus

$$c_i = \bigcup_{\gamma<\lambda} \pi(R_\gamma)\,.$$

Case 1. $\exists\gamma < \lambda(b_i = R_\gamma)$; i.e. R_γ is eventually constant as $\gamma \to \lambda$.

PROOF. Let $b_i = R_\gamma = \bigcup_{\eta<\lambda} R_\eta$. Then $c_i = \bigcup_{\eta<\lambda} \pi(R_\eta) = \pi(R_\gamma) = \pi(b_i)$. Thus $c_i \in |\eta|$.

Case 2. Otherwise.

PROOF. For $x \in b_i$, let $\gamma_x =$ least γ such that $x \in R_\gamma$ and if $y \in b_i$, $x \le y$ iff $\gamma_x \le \gamma_y$. Thus $\le$ is a prewellorder of b_i of limit order type. Notice $\le$ is computable within $\mathcal{M}$ from b_i, so that our solidity hypothesis on r means $b_i \in |\mathcal{M}|$ and $\le\, \in |\mathcal{M}|$. Now clearly, for $x \in b_i$

$$R_{\gamma_x} = \{y \in b_i \mid y \le x\}$$

so

$$\pi(R_{\gamma_x}) = \{y \in \pi(b_i) \mid y \le^* \pi(x)\}$$

where $\le^* = \pi(\le)$. By case hypothesis

$$c_i = \{y \in \pi(b_i) \mid \exists x \in b_i(y \le^* \pi(x))\}\,,$$

so that c_i is an $\le^*$ initial segment of $\pi(b_i)$.

Subcase 2A. $\mathrm{cof}^{\mathcal{M}}(\le) \ne \mathrm{crit}(E)$. In this case, ran π is $\le^*$ cofinal in $\pi(b_i)$, so that $c_i = \pi(b_i)$, and $c_i \in |\mathcal{N}|$.

Subcase 2B. $\text{cof}^{\mathcal{M}}(\leq) = \text{crit}(E)$. Let $\kappa = \text{crit}(E)$, $f \in |\mathcal{M}|$, $f : \kappa \to b_i$ such that ran f is $\leq$-cofinal. Then

$$\begin{aligned} y \in c_i &\Leftrightarrow \exists\alpha < \kappa(y \leq^* \pi(f(\alpha))) \\ &\Leftrightarrow \exists\alpha < \kappa(y \leq^* \pi(f)(\alpha)) \end{aligned}$$

so that $c_i \in |\mathcal{N}|$, as desired.

□

Of course, we shall need to know that fine structure "up to level n" is preserved not just under passage to Ult_n, but under iteration of this process. The following lemma summarizes the important facts.

Lemma 4.7. *Let $\mathcal{M} = \mathcal{M}_0$ be n-sound, where $n \leq \omega$. Suppose that for $\alpha < \theta$,*

$$\mathcal{M}_{\alpha+1} = \text{Ult}_n(\mathcal{M}_\alpha, E_\alpha),$$

where E_α is close to $\mathcal{M}_\alpha$, and

$$\mathcal{M}_\lambda = \operatorname{dir\,lim}_{\beta<\lambda} \mathcal{M}_\beta$$

for $\lambda \leq \theta$ a limit. (We assume each $\mathcal{M}_\alpha$ is wellfounded.) Let $\pi_{0\theta} : \mathcal{M} \to \mathcal{M}_\theta$ be the canonical embedding. Then

(a) $\pi_{0\theta}$ is an n-embedding.

If, in addition, $\mathcal{M}$ is $n+1$-sound (so $n < \omega$) and $\rho_{n+1}^{\mathcal{M}} \leq \text{crit}\ \pi_{0\theta}$, then

(b) $\rho_{n+1}^{\mathcal{M}} = \rho_{n+1}^{\mathcal{M}_\theta}$.
(c) $\pi_{0\theta}(p_{n+1}(\mathcal{M})) = p_{n+1}(\mathcal{M}_\theta$.
(d) $\mathcal{M}_\theta$ is $n+1$-solid, and in fact $\mathfrak{C}_{n+1}(\mathcal{M}_\theta) = \mathcal{M}$, and letting

$$\sigma : \mathfrak{C}_{n+1}(\mathcal{M}_\theta) \to \mathfrak{C}_n(\mathcal{M}_\theta) = \mathcal{M}_\theta$$

be the inverse of the collapse, $\sigma = \pi_{0\theta}$.

Proof. This is a fairly routine induction on θ, using Lemmas 4.4, 4.5, and 4.6. The successor case is immediate from these lemmas, so let θ be a limit. Then (a), (b) are obvious, and (d) follows easily from (c). We sketch a proof of (c): let $p_{n+1}(\mathcal{M}) = \langle \bar{r}, \bar{u} \rangle$, where $\bar{r} = \langle \alpha_0 \cdots \alpha_\ell \rangle$. For $\gamma < \theta$ let

$$b_i^\gamma = \text{Th}_{n+1}^{\mathcal{M}_\gamma}(\pi_{0\gamma}(\alpha_i) \cup \{\pi_{0\gamma}(\alpha_1), \ldots, \pi_{0\gamma}(\alpha_{i-1}), \pi_{0\gamma}(\bar{u})\}).$$

Part of our induction hypothesis should be that $b_i^\gamma \in |\mathcal{M}_\gamma|$ for $0 \leq i \leq \ell$. This follows from 4.6 for successor θ, and for limit θ, our current case, from the proof of 4.6. For that proof shows that for each fixed i there are most finitely many

$\gamma < \theta$ such that $\pi_{\gamma+1}(b_i^\gamma) \neq b_i^{\gamma+1}$, since this can occur only when the ordering $\leq_i^\gamma$ of that proof has $\mathcal{M}_\gamma$-cofinality equal to crit E_γ, but when that happens $\leq_i^{\gamma+1}$ also has $\mathcal{M}_{\gamma+1}$-cofinality crit $E_\gamma <$ crit $E_{\gamma+1}$. Thus we can find $\gamma < \theta$ such that for all i, and all η such that $\gamma < \eta < \theta$, $\pi_{\gamma\eta}(b_i^\gamma) = b_i^\eta$.

One can now easily check that $b_i^\theta = \pi_{\gamma\theta}(b_i^\gamma)$ for all i. This in turn implies that $\pi_{0\theta}(\bar{r})$ is the $n+1$st standard parameter of $(\mathcal{M}_\theta, \pi_{0\theta}(\bar{u}))$. The rest of (c) is easy. □

REMARK. Under the hypotheses of 4.7 (including that $\mathcal{M}$ is $n+1$-sound and $\rho_{n+1}^{\mathcal{M}} \leq \text{crit}(\pi_{0\theta})$), we see that $\mathcal{M}_\theta$ is not $n+1$-sound. If $\text{crit}(i_{0\theta}) \neq \tau^{\mathcal{M}_0}[\bar{\alpha}, \pi_{0\theta}(p_{n+1}(\mathcal{M}))]$ then $\forall\bar{\alpha} \in [\rho_{n+1}^{\mathcal{M}}]^{<\omega}$ and $\tau \in \text{Sk}_{n+1}$.

We generalize the key tool of Martin-Steel [MS] to the fine structure context.

DEFINITION 5.0.1. A *tree order on* α (for $\alpha \in \mathrm{OR}$) is a strict partial order T of α such that

(1) $\beta \neq 0 \Rightarrow 0T_\beta$,
(2) $\beta T\gamma \Rightarrow \beta < \gamma$,
(3) $\{\beta \mid \beta T\gamma\}$ is wellordered by T,
(4) γ limit $\Rightarrow \{\beta \mid \beta T\gamma\}$ is cofinal in γ (i.e. $\in$ cofinal) and
(5) γ successor $\Leftrightarrow \gamma$ is a T-successor.

DEFINITION 5.0.2. If T is a tree order then

$$[\beta,\gamma]_T = \{\eta \mid \eta = \beta \vee \beta T\eta T\gamma \vee \eta = \gamma\}$$

and similarly for $(\beta,\gamma]_T$, $[\beta,\gamma)_T$, and $(\beta,\gamma)_T$.

DEFINITION 5.0.3. T-Pred$(\gamma+1)$ is the unique ordinal $\eta T\gamma$ such that $(\eta,\gamma)_T = \varnothing$.

DEFINITION 5.0.4. Let $\mathcal{M} = \mathcal{J}_\beta^{\vec{E}}$ be a ppm. Then for $\gamma \leq \beta$, $\mathcal{J}_\gamma^{\mathcal{M}} = \mathcal{J}_\gamma^{\vec{E}}$. For $\gamma > \beta$, $\mathcal{J}_\gamma^{\mathcal{M}}$ is undefined.

DEFINITION 5.0.5. Let $\mathcal{M}$ and $\mathcal{N}$ be ppm's. Then $\mathcal{M}$ is an *initial segment* of $\mathcal{N}$ iff $\exists\gamma(\mathcal{M} = \mathcal{J}_\gamma^{\mathcal{N}})$. $\mathcal{M}$ is a *proper initial segment* of $\mathcal{N}$ iff $\mathcal{M}$ is an initial segment of $\mathcal{N}$ and $\mathcal{N}$ is not an initial segment of $\mathcal{M}$.

Notice that if $\beta \in \mathrm{dom}\ \vec{E}$, then $(J_\beta^{\vec{E}}, \in, \vec{E} \restriction \beta)$ is not an initial segment of $\mathcal{J}_\beta^{\vec{E}}$ according to our definition, although we might reasonably have regarded it as such.

DEFINITION 5.0.6. Let $\mathcal{M}$ and $\mathcal{N}$ be ppm's. Then $\mathcal{M}$ and $\mathcal{N}$ *agree below* γ iff $\mathcal{J}_\beta^{\mathcal{M}} = \mathcal{J}_\beta^{\mathcal{N}}$ for all $\beta < \gamma$. (In particular, $\mathcal{J}_\beta^{\mathcal{M}}$ is defined iff $\mathcal{J}_\beta^{\mathcal{N}}$ is defined, for all $\beta < \gamma$.)

If $\mathcal{M}$ is a ppm then a *iteration tree of length* θ *on* $\mathcal{M}$ is a 4-tuple

$$\mathcal{T} = \langle T, \deg, D, \langle E_\alpha, \mathcal{M}^*_{\alpha+1} \mid \alpha+1 < \theta\rangle\rangle,$$

where T is a tree order, which satisfies conditions (1–8) below. We write ρ_α for the natural length of E_α. We will also define ppm $\mathcal{M}_\alpha$ for $\alpha < \theta$ and embeddings $i_{\alpha,\beta}\colon \mathcal{M}_\alpha \to \mathcal{M}_\beta$ for ordinals α and β less than θ such that $\alpha T\beta$ and $D \cap (\alpha\beta]_T = \varnothing$.

(1) $\mathcal{M}_0 = \mathcal{M}$, and each $\mathcal{M}_\alpha$ is a ppm.

(2) E_α is the extender coded by $\dot{F}^{\mathcal{N}}$, for some active ppm $\mathcal{N}$ which is an initial segment of $\mathcal{M}_\alpha$.

(3) $\alpha < \beta \Rightarrow \mathrm{lh}(E_\alpha) < \mathrm{lh}(E_\beta)$.

(4) If T-Pred$(\alpha+1) = \beta$ then $\kappa = \mathrm{crit}\, E_\alpha < \rho_\beta$, and $\mathcal{M}^*_{\alpha+1}$ is an initial segment $\mathcal{J}_\gamma^{\mathcal{M}_\beta}$ of $\mathcal{M}_\beta$ such that $P(\kappa) \cap \mathcal{M}^*_{\alpha+1} = P(\kappa) \cap \mathcal{N}$. Moreover

$$\alpha + 1 \in D \Leftrightarrow \mathcal{J}_\gamma^{\mathcal{M}_\beta} \text{ is a proper initial segment of } \mathcal{M}_\beta .$$

If we take $n = \deg(\alpha+1)$ then $\kappa < \rho_n^{\mathcal{M}^*_{\alpha+1}}$ and

$$\mathcal{M}_{\alpha+1} = \mathrm{Ult}_n(\mathcal{M}^*_{\alpha+1}, E_\alpha)$$

and if $\alpha + 1 \notin D$, then

$$i_{\beta,\alpha+1} = \text{canonical embedding of } \mathcal{M}_\beta \text{ into } \mathrm{Ult}_n(\mathcal{M}_\beta, E_\alpha),$$

and $i_{\gamma,\alpha+1} = i_{\beta,\alpha+1} \circ i_{\gamma,\beta}$ for all $\gamma T \beta$ such that $(\gamma, \beta]_T \cap D = \varnothing$.

(5) If $\lambda < \theta$ is a limit, then $D \cap [0, \lambda)_T$ is finite, and letting γ be the largest element of $D \cap [0, \lambda)_T$,

$$\mathcal{M}_\lambda = \text{direct limit of } \mathcal{M}_\alpha, \alpha \in [\gamma, \lambda)_T, \text{ under the } i_{\alpha\beta}\text{'s}$$
$$i_{\eta\lambda} = \text{canonical embedding of } \mathcal{M}_\eta \text{ into } \mathcal{M}_\lambda, \text{ for } \eta \in [\gamma, \lambda)_T .$$

(6) $\mathcal{M}^*_{\alpha+1}$ is $\deg(\alpha+1)$-sound.

(7) If $\gamma + 1 T \alpha + 1$ and $D \cap (\gamma+1, \alpha+1]_T = \varnothing$, then $\deg(\gamma+1) \geq \deg(\alpha+1)$.

(8) For $\lambda \leq \theta$ a limit, $\deg(\lambda) = \deg(\alpha+1)$, for all sufficiently large $\alpha + 1\, T \lambda$.

Notice that $\mathcal{T}$ determines the ordinals ρ_α's, the embeddings $i_{\alpha\beta}$'s, and the ppm $\mathcal{M}_\alpha$.

Conditions (6–8) can be dropped in some contexts. Condition (6) guarantees that $i^*_{\alpha+1}$ is a $\deg(\alpha+1)$-embedding. Condition (7) says that the ultrapowers taken along branches of $\mathcal{T}$ are of decreasing elementarity; it allows us to "copy $\mathcal{T}$" via certain embeddings.

Lemma 5.1. *Let $\mathcal{T} = \langle T, \deg, D, \langle E_\alpha, \mathcal{M}^*_{\alpha+1} \mid \alpha + 1 < \theta \rangle\rangle$ be an iteration trec. Then if $\alpha < \beta < \theta$*

(1) *$\mathcal{M}_\alpha$, and $\mathcal{M}_\beta$ agree below $\mathrm{lh}\, E_\alpha$, and*
(2) *$\mathrm{lh}(E_\alpha)$ is a cardinal of $\mathcal{M}_\beta$, and in particular $\mathcal{M}_\alpha$ and $\mathcal{M}_\beta$ do not agree below $\mathrm{lh}(E_\alpha) + 1$.*

Proof. By induction on β. Let $\beta = \gamma + 1$. Since $\alpha \leq \gamma \Rightarrow \mathrm{lh}\, E_\alpha \leq \mathrm{lh}\, E_\gamma$, it is enough for (1) to show that $\mathcal{M}_{\gamma+1}$ and $\mathcal{M}_\gamma$ agree below $\mathrm{lh}\, E_\gamma$. Let $E_\gamma = \dot{F}^{\mathcal{N}}$

where $\mathcal{N}$ is an initial segment of $\mathcal{M}_\gamma$. Now $\mathcal{M}_\gamma$ agrees with $\mathrm{Ult}_0(\mathcal{N}, E_\gamma)$ below $\mathrm{lh}\, E_\gamma$ by coherence. But $M_{\gamma+1} = \mathrm{Ult}_n(\mathcal{M}^*_{\gamma+1}, E_\gamma)$, where $\mathcal{M}^*_{\gamma+1}$ is an initial segment of $\mathcal{M}_\delta$, some $\delta \leq \gamma$, with crit $E_\gamma < \min(\mathrm{OR}^{\mathcal{M}^*_{\gamma+1}}, \mathrm{lh}\, E_\delta)$. By induction, $\mathcal{M}_\delta$ agrees with $\mathcal{M}_\gamma$ below $\mathrm{lh}\, E_\delta$, hence below crit E_γ. Thus $\mathcal{M}^*_{\gamma+1}$ agrees with $\mathcal{M}_\gamma$ below crit E_γ. So $\mathcal{M}_{\gamma+1}$ agrees with $\mathrm{Ult}_0(\mathcal{N}, E_\gamma)$ below $\mathrm{lh}\, E_\gamma$, hence with $\mathcal{M}_\gamma$ below $\mathrm{lh}\, E_\gamma$. (Notice here that if $\eta < \mathrm{lh}\, E_\gamma$, then the function representing $\mathcal{J}_\eta^{\mathcal{M}_{\gamma+1}}$ is in both $\mathcal{M}^*_{\gamma+1}$ and $\mathcal{N}$. In fact, $P(\text{crit } E_\gamma) \cap \mathcal{M}^*_{\gamma+1} = P(\text{crit } E_\gamma) \cap \mathcal{N}$. For $\subseteq$ is true by fiat and $\supseteq$ by our induction hypotheses.)

For the second assertion it is enough to show $\mathrm{lh}\, E_\gamma$ is a cardinal in $\mathcal{M}_{\gamma+1}$ (using (1) and strong acceptability). Let us adopt the notation of the last paragraph. The definition of good extender sequence guarantees $\mathrm{lh}\, E_\gamma$ is a cardinal in $\mathrm{Ult}_0(\mathcal{N}, E_\gamma)$. But if $A \subseteq \mathrm{lh}\, E_\gamma$ and $A \in \mathcal{M}_{\gamma+1}$ then $A = [a, f]$ for some function $f : [\mathrm{crit}(E_\gamma)]^n \to \mathcal{J}^{\mathcal{M}^*_{\gamma+1}}_{\mathrm{crit}(E_\gamma)}$ in $\mathcal{M}^*_{\gamma+1}$. But then $f \in \mathcal{N}$, so $A \in \mathrm{Ult}_0(\mathcal{N}, E_\gamma)$, so A doesn't collapse $\mathrm{lh}\, E_\gamma$.

We leave the case β is a limit to the reader. □

Let H_λ be the set of sets hereditarily of cardinality $< \lambda$. From 5.1 we get, using the notation there, that if $\alpha < \beta$ and $\lambda = \mathrm{lh}\, E_\alpha$, then $H_\lambda^{\mathcal{M}_\beta} = |\mathcal{J}_\lambda^{\mathcal{M}_\alpha}|$.

A few miscellaneous remarks on the definition of an iteration tree:

(a) It is easy to see from the above that if $\mathcal{T}$ is an iteration tree of length θ, $\alpha < \beta < \theta$, and F is an extender from the $\mathcal{M}_\beta$ sequence (i.e. F on $\dot{E}^{\mathcal{M}_\beta}$ or $F = \dot{F}^{\mathcal{M}_\beta}$), then $E_\alpha \restriction \rho_\alpha \neq F \restriction \rho_\alpha$. For suppose $E_\alpha \restriction \rho_\alpha = F \restriction \rho_\alpha$. If F is on $\dot{E}^{\mathcal{M}_\beta}$, this implies $E_\alpha \restriction \rho_\alpha \in \mathcal{M}_\beta$, and therefore that $\mathrm{lh}\, E_\alpha$ is not a cardinal of $\mathcal{M}_\beta$, contrary to 5.1. If $F = \dot{F}^{\mathcal{M}_\beta}$, then $\dot{\nu}^{\mathcal{M}_\beta} = \nu \geq \mathrm{lh}\, E_\alpha$ since $\mathrm{lh}\, E_\alpha$ is a cardinal of $\mathcal{M}_\beta$, and $\rho_\alpha < \nu$. By the initial segment condition on good extender sequences, $F \restriction \rho_\alpha \in \mathcal{M}_\beta$. Since $E_\alpha \restriction \rho_\alpha$ collapses $\mathrm{lh}\, E_\alpha$, we again have a contradiction.

(b) The demand in (4) that crit $E_\alpha < \rho_\beta$, rather than just crit $E_\alpha < \mathrm{lh}\, E_\beta$, makes a difference only when $E_\beta = \dot{F}^{\mathcal{P}}$ for some $\mathcal{P}$ of type III, so that $\rho_\beta = \nu^{\mathcal{P}}$, and crit $E_\alpha = \rho_\beta = \nu^{\mathcal{P}}$. In this case our official definition won't allow us to apply E_α to an initial segment of $\mathcal{M}_\beta$ to form $\mathcal{M}_{\alpha+1}$.

(c) Suppose we have an iteration tree

$$\mathcal{T} = \langle T, \deg, D, \langle E_\gamma, \mathcal{M}^*_{\gamma+1} \mid \gamma + 1 < \alpha + 1\rangle\rangle ,$$

so that the last model $\mathcal{M}_\alpha$ of $\mathcal{T}$ is determined. Suppose $F = \dot{F}^{\mathcal{P}}$ for some initial segment $\mathcal{P}$ of $\mathcal{M}_\alpha$. How may we extend $\mathcal{T}$ one step further so that $F = E_\alpha$? Let us assume all ultrapowers to follow are wellfounded. Assume also that $\mathrm{lh}\, F > \mathrm{lh}\, E_\gamma$ for all $\gamma < \alpha$. Let $\kappa = \text{crit } F$.

(i) We may set $\alpha T \alpha + 1$ and take $\mathcal{M}^*_{\alpha+1}$ to be any initial segment of $\mathcal{M}_\alpha$ such that $\mathcal{P}$ is an initial segment of $\mathcal{M}^*_{\alpha+1}$ and $P(\kappa) \cap |\mathcal{P}| = P(\kappa) \cap |\mathcal{M}^*_{\alpha+1}|$. Notice that if Q is a type III initial segment of $\mathcal{M}_\alpha$, $\mathcal{P}$ an initial segment of Q, and $P(\kappa)^{\mathcal{P}} = P(\kappa)^{Q}$, then $\kappa < \nu^{Q}$ since $(\kappa^+)^{\mathcal{P}} = (\kappa^+)^{Q}$, whereas ν^{Q} is the largest cardinal of Q. Thus we can form $\mathrm{Ult}(Q^{\mathrm{sq}}, F)$.

(ii) Suppose $\beta < \alpha$ and $\kappa < \rho_\beta$. Then we may set $\beta = T\text{-pred}(\alpha+1)$. The candidates for $\mathcal{M}^*_{\alpha+1}$ are precisely those structures $\mathcal{J}_\gamma^{\mathcal{M}_\beta}$ such that $\gamma \geq \mathrm{lh}\, E_\beta$ and $P(\kappa) \cap |\mathcal{J}_\gamma^{\mathcal{M}_\beta}| = P(\kappa) \cap |\mathcal{J}_{\mathrm{lh}\, E_\beta}^{\mathcal{M}_\beta}|$. Any of these candidates will do for $\mathcal{M}^*_{\alpha+1}$. Notice again that if $Q = \mathcal{J}_\gamma^{\mathcal{M}_\beta}$ for such a γ, then $\kappa < \nu^{Q}$ as $(\kappa^+)^{\mathcal{P}}$ is a cardinal of Q. So we can squash Q if necessary and still apply F.

In almost all of the iteration trees used in this paper, the extension of $\mathcal{T}$ to $\alpha+1$ will be determined by the choice of E_α. We take $T\text{-Pred}(\alpha+1)$ to be the least ordinal α^*, if there is one, such that $\rho_{\alpha^*} > \mathrm{crit}(E_\alpha)$ and $\alpha^* = T\text{-Pred}(\alpha+1) = \alpha$ otherwise. Then we take $\mathcal{M}^*_{\alpha+1}$ to be the largest initial segment of $\mathcal{M}_{\alpha^*}$ which does not contain any subset of $\mathrm{crit}(E_\alpha)$ other than those measured by E_α. Finally we take $\deg(\alpha+1)$ to be the the largest ordinal such that $\mathrm{crit}(E_\alpha) < \rho_n^{\mathcal{M}^*_{\alpha+1}}$. See the definition of n-maximal, definition 6.1.2, for details.

Iterability. If T is a tree order on θ, then a branch of T is a set $b \subset \theta$ such that b is wellordered by T with limit order type, and $\forall\alpha \in b \forall\beta(\beta T \alpha \Rightarrow \beta \in b)$. We call b *cofinal* iff $\sup b = \theta$. We call b *maximal* iff $b \neq [0,\lambda)_T$ for all $\lambda < \theta$. If $\mathcal{T} = \langle T, \deg, D, \langle E_\alpha, \mathcal{M}^*_{\alpha+1} \mid \alpha+1 < \theta\rangle\rangle$ is an iteration tree, then a (maximal, cofinal) branch of $\mathcal{T}$ is a (maximal, cofinal) branch of T. If b is a branch of T such that $D \cap b$ is finite, with largest element γ, then we set

$$\mathcal{M}_b = \text{direct limit of } \mathcal{M}_\alpha,\ \alpha \in b - \gamma, \quad \text{under the} \quad i_{\alpha\beta}\text{'s}.$$

We say a *branch* b of $\mathcal{T}$ *is wellfounded* iff $D \cap b$ is finite and $\mathcal{M}_b$ is wellfounded.

We now state the iterability property which qualifies premice having no more than one Woodin cardinal as mice. We shall eventually show that all levels of the model we construct have this property by quoting results of Martin-Steel [MS].

DEFINITION 5.1.1. If $\mathcal{T} = \langle T, \deg, D, \langle E_\alpha, \mathcal{M}^*_{\alpha+1} \mid \alpha+1 < \theta\rangle\rangle$ then for $\beta \leq \theta$

$$\mathcal{T} \restriction \beta = \langle T \cap (\beta \times \beta), \deg \restriction \beta, D \cap \beta, \langle E_\alpha, \mathcal{M}^*_{\alpha+1} \mid \alpha+1 < \beta\rangle\rangle\,.$$

DEFINITION 5.1.2. Let $\mathcal{T}$ be an iteration tree of length θ. $\mathcal{T}$ is *simple* if and only if every maximal wellfounded branch of $\mathcal{T}$ is cofinal in θ, and $\mathcal{T}$ has at most one cofinal in θ wellfounded branch.

Notice that by definition 5.0.1(4) it follows that $\mathcal{T}$ is simple iff for every limit $\lambda \leq \theta$, $\mathcal{T} \restriction \lambda$ has at most one cofinal wellfounded branch.

We shall deal almost exclusively with simple iteration trees. The fact that it suffices to do so is one of the key things we must prove. (c.f. Theorem 6.2.)

DEFINITION 5.1.3. Let $\kappa \leq \omega$. Then an iteration tree $\mathcal{T}$ is *k-bounded* iff $\deg^{\mathcal{T}}(\alpha+1) \leq k$ whenever α is such that $[0, \alpha+1]_T \cap D^{\mathcal{T}} = \varnothing$.

Notice that by clause (7) in the definition of "iteration tree", if $\deg(\alpha+1) \leq k$ whenever $\alpha+1 \notin D$ and T-pred$(\alpha+1) = 0$, then $\mathcal{T}$ is k-bounded.

DEFINITION 5.1.4. Let $\mathcal{M}$ be a ppm, and let $k \leq \omega$. (1) $\mathcal{M}$ is *singly k-iterable* if any k-bounded iteration tree

$$\mathcal{T} = \langle T, \deg, D, \langle E_\alpha, \mathcal{M}^*_{\alpha+1} \mid \alpha+1 < \theta \rangle \rangle$$

such that $\mathcal{T} \restriction \lambda$ is simple for all $\lambda < \theta$ satisfies conditions (a) and (b) below:

(a) If θ is a limit ordinal, then $\mathcal{T}$ has a cofinal wellfounded branch.

(b) Suppose $\alpha < \theta = \beta + 1$ and $\mathcal{N}$ is an active initial segment of $\mathcal{M}_\beta$, such that $\operatorname{crit}(\dot{F}^{\mathcal{N}}) < \rho_\alpha$, and suppose that $\mathcal{P} = \mathcal{J}^{\mathcal{M}_\alpha}_\gamma$ for some $\gamma \geq \operatorname{lh} E_\alpha$, with $\kappa = \operatorname{crit}(\dot{F}^{\mathcal{N}}) < \rho^{\mathcal{P}}_n$ and $P(\kappa) \cap |\mathcal{P}| \subseteq \mathcal{N}$. Then

$$\operatorname{Ult}_n(\mathcal{P}, \dot{F}^{\mathcal{N}}) \text{ is wellfounded}$$

(provided also $n \leq k$ when $[0, \alpha]_T \cap D \neq \varnothing$ and $\mathcal{P} = \mathcal{M}_\alpha$).

(2)We say $\mathcal{M}$ is *k-iterable* if it is singly k-iterable and satisfies conditions (a) and (b) below:

(a) If $n < \omega$, and $\langle \mathcal{T}_i : i \leq n \rangle$ is a sequence of iteration trees such that $\mathcal{T}_0$ is a k-bounded simple iteration tree on $\mathcal{M}$, and for $i > 0$ $\mathcal{T}_i$ is a simple iteration tree on the last model $\mathcal{M}^{\mathcal{T}_{i-1}}_{\theta_{i-1}}$ of $\mathcal{T}_{i-1}$, and $\mathcal{T}_i$ is k-bounded whenever $D^{\mathcal{T}_j} \cap [0, \theta_j]_{T_j} = \varnothing$ for all $j < i$, then the last model $\mathcal{M}^{\mathcal{T}_n}_{\theta_n}$ of $\mathcal{T}_n$ is singly k-iterable.

(b) Suppose that $\langle \mathcal{T}_i : i < \omega \rangle$ is as in (a). Then $[0, \theta_i]_{T_i} \cap D_i = \varnothing$ for all but finitely many i, so that we have a canonical embedding $\tau_i : \mathcal{M}^i_0 \to \mathcal{M}^{i+1}_0 = \mathcal{M}^i_{\theta_i}$ defined for sufficiently large $i < \omega$. Moreover, the direct limit of the $\mathcal{M}^i_0$'s under the τ_i's is wellfounded.

It is easy to see that if $\mathcal{M}$ is k-iterable, $\mathcal{T}$ is a k-bounded simple tree on $\mathcal{M}$, and $\mathcal{P}$ is a model on $\mathcal{T}$, then $\mathcal{P}$ is k-iterable.

It may seem that we can derive (2) and (3) from (1). Given $\mathcal{T}_i$'s as in (2) or (3), we can lay the $\mathcal{T}_i$'s "end-to-end" and produce a tree $\mathcal{S}$ to which we can then apply (1). The problem is that $\mathcal{S}$ may not be, formally speaking, an iteration tree: we may have $\alpha < \beta$ such that $\operatorname{lh} E^{\mathcal{S}}_\alpha \not< \operatorname{lh} E^{\mathcal{S}}_\beta$. This can definitely occur in the proof of the Dodd-Jensen lemma on the minimality of iteration maps, which is our application of (2) and (3). Rather than generalize the definition of "iteration tree" we prefer to complicate the definition of iterability.

The k-iterability of $\mathcal{M}$ allows us to build k-bounded iteration trees on $\mathcal{M}$ freely as long as the tree built so far is simple. For then (1b) guarantees we can proceed at successor steps without fear of illfoundedness. Clause (1a) guarantees that at a limit ordinal λ we have a cofinal in λ wellfounded branch. Thus we can choose this branch to be $[0, \lambda)_T$.

It should be remarked that a theorem of Woodin asserts that the model $L[\vec{E}]$ which we are constructing is not fully iterable, in the sense that there is a tree which is a member of $L[\vec{E}]$ but which has no well founded branch which is a member of $L[\vec{E}]$. If we make the additional assumption that every set has a sharp then we can prove that $V \models L[\vec{E}]$ is iterable: that is, every tree on $L[\vec{E}]$ has a well founded branch, with both the tree and the branch being in V. It is a theorem of ZFC that every iteration tree which involves only extenders from a proper initial segment of the sequence $\vec{E}$ has a well founded branch, so that this much iterablity is true in both V and $L[\vec{E}]$. The proof that our construction works will depend on this iterability in V of initial segments of $\vec{E}$. It is important for this that $L[\vec{E}]$ has no more than the one Woodin cardinal, which is the supremum of $\mathrm{dom}(\vec{E})$.

DEFINITION 5.1.5. Let $\mathcal{M}$ be a ppm. Then $\mathcal{M}$ is 1-*small* iff whenever $\kappa = \mathrm{crit}\, \dot{F}^{\mathcal{N}}$ for some initial segment $\mathcal{N}$ of $\mathcal{M}$, then $\mathcal{J}^{\mathcal{M}}_{\kappa} \models$ "There are no Woodin cardinals".

It is possible for a 1-small ppm $\mathcal{M}$ to satisfy "there is a Woodin cardinal"; however, such an $\mathcal{M}$ cannot satisfy "there is a sharp for an inner model with a Woodin cardinal".

DEFINITION 5.1.6. A 1-*small mouse* is a 1-small, ω-iterable premouse.

DEFINITION 5.1.7. A 1-*small coremouse* is a 1-small mouse which is completely sound.

In general (for models with more than a Woodin cardinal) ω-iterability will not convert a premouse into a mouse.

Since all the mice we shall deal with in the moderately near future will be 1-small, we make the temporary convention:

$$\text{mouse} = \text{1- small mouse}$$

$$\text{coremouse} = \text{1- small coremouse}$$

Embeddings of Iteration Trees. We now head toward the Dodd-Jensen lemma on the minimality of iteration maps. For that we must show, given an embedding $\pi: \mathcal{M} \to \mathcal{N}$ and a iteration tree $\mathcal{T}$ on $\mathcal{M}$, how to extend π to an embedding from $\mathcal{T}$ into an iteration tree $\mathcal{U}$ on $\mathcal{N}$. Since not all of the embeddings involved will be full n-embeddings we need a new definition:

DEFINITION. We say $\pi: \mathcal{M} \to \mathcal{N}$ is a *weak n-embedding* if $\mathcal{M}$ and $\mathcal{N}$ are premice of types I or II or sppm's, and there is a set $X \subset \mathcal{M}$ such that the following four conditions hold:

(i) The models $\mathcal{M}$ and $\mathcal{N}$ are n-sound, and X is a subset of $\mathcal{M}$ such that

$\{p_n^{\mathcal{M}}, \rho_n^{\mathcal{M}}\} \subset X$, and X is cofinal in $\rho_n^{\mathcal{M}}$.

(ii) π is $r\Sigma_n$ (respectively $q\,\Sigma_n$) elementary, and π is $r\Sigma_{n+1}$ (respectively $q\,\Sigma_{n+1}$) elementary on parameters from X.

(iii) $\pi(p_i(\mathcal{M})) = p_i(\mathcal{N})$ for $i \leq n$

(iv) $\pi(\rho_i(\mathcal{M})) = \rho_i(\mathcal{N})$ for $i < n$, and $\sup \pi''\rho_n(\mathcal{M}) \leq \rho_n(\mathcal{N})$.

If $\mathcal{M}$ and $\mathcal{N}$ are type III, a weak n-embedding from $\mathcal{M}$ to $\mathcal{N}$ is a weak n-embedding from $\mathcal{M}^{sq}$ to $\mathcal{N}^{sq}$.)

Note that this definition is obtained from the definition of a n-embedding by weakening clause (ii) from $r\Sigma_{n+1}$ to $r\Sigma_n$ except for parameters from X, and weakening clause (iv) by eliminating the requirement that $\pi''\rho_n(\mathcal{M})$ be cofinal in $\rho_n(\mathcal{N})$. Normally it is the existence of a set X which is important, rather than the choice of the set X.

The following is a useful fact about (n, X)-embeddings:

Proposition. *Suppose that $\pi\colon \mathcal{P} \to \mathcal{Q}$ is a weak n-embedding and κ is an ordinal in $\mathrm{OR}^{\mathcal{P}}$. Then $\mathcal{P} \models \kappa$ is a cardinal if and only if $\mathcal{Q} \models \pi(\kappa)$ is a cardinal.*

PROOF. This a is obvious if $n \geq 1$, so let $n = 0$. Recall $\rho_0(\mathcal{P}) = \mathrm{OR}^{\mathcal{P}}$, so that the set X on which π is $r\Sigma_1$ elementary is cofinal in $\mathrm{OR}^{\mathcal{P}}$. Fix κ s.t. $\mathcal{P} \models \kappa$ is a cardinal, and let $\mu \in X$ be such that $\kappa < \mu$. Let $\xi \in X$, $\mu < \xi$, be such that

$$\mathrm{card}^{\mathcal{P}}(\mu) = \mathrm{card}^{S_\xi^{\mathcal{P}}}(\mu),$$

where "$S_\xi^{\mathcal{P}}$" refers to the ξth level of the Jensen S-hierarchy. Then

$$\mathcal{P} \models \mathrm{card}^{S_\xi^{\mathcal{P}}}(\mu) \text{ is a cardinal}$$

and as ξ, $\mu \in X$

$$\mathcal{Q} \models \mathrm{card}^{S_{\pi(\xi)}^{\mathcal{Q}}}(\pi(\mu)) \text{ is a cardinal.}$$

So, setting $\nu = \mathrm{card}^{S_\xi^{\mathcal{P}}}(\mu)$, we know that $\kappa \leq \nu$ and $\pi(\nu)$ is a cardinal of $\mathcal{Q}$. If $\kappa = \nu$ we're done. If $\kappa < \nu$, then $\mathcal{J}_\nu^{\mathcal{P}} \models \kappa$ is a cardinal, so since the relation $R(z, x) \Leftrightarrow$ "x is a cardinal relative to z" is Σ_0-in-$\mathcal{L} \setminus \{\dot{F}\}$ we know that $\mathcal{J}_{\pi(\nu)}^{\mathcal{Q}} \models \pi(\kappa)$ is a cardinal, and hence $\mathcal{Q} \models \pi(\kappa)$ is a cardinal. □

Lemma 5.2 (Shift lemma). *Let $\bar{\mathcal{M}}$ and $\bar{\mathcal{N}}$ be ppm's, let $\bar{\kappa} = \mathrm{crit}(\dot{F}^{\bar{\mathcal{N}}})$, and let*

$$\pi : \bar{\mathcal{M}} \to \mathcal{M} \quad \text{be an weak } n\text{-embedding } (n \leq \omega) \text{ and}$$

$$\psi : \bar{\mathcal{N}} \to \mathcal{N} \quad \text{be a weak 0-embedding}$$

such that $\bar{\mathcal{M}}$ and $\bar{\mathcal{N}}$ agree below $(\bar{\kappa}^+)^{\bar{\mathcal{M}}} \leq (\bar{\kappa}^+)^{\bar{\mathcal{N}}}$, while $\mathcal{M}$ and $\mathcal{N}$ agree below $(\kappa^+)^{\mathcal{M}} \leq (\kappa^+)^{\mathcal{N}}$, and $\pi \restriction (\bar{\kappa}^+)^{\bar{\mathcal{M}}} = \psi \restriction (\bar{\kappa}^+)^{\bar{\mathcal{M}}}$. Suppose $\bar{\kappa} < \rho_n^{\bar{\mathcal{M}}}$, so *that* $\mathrm{Ult}_n(\bar{\mathcal{M}}, \dot{F}^{\bar{\mathcal{N}}})$ *makes sense, as does* $\mathrm{Ult}_n(\mathcal{M}, \dot{F}^{\mathcal{N}})$, *and that both of these ultrapowers are wellfounded. Then there is an embedding* $\sigma\colon \mathrm{Ult}_n(\bar{\mathcal{M}}, \dot{F}^{\bar{\mathcal{N}}}) \to \mathrm{Ult}_n(\mathcal{M}, \dot{F}^{\mathcal{N}})$ *satisfying the following four conditions:*

(a) *The map* σ *is an weak* n*-embedding, and if* π *is an* n*-embedding then so is* σ.
(b) $\mathrm{Ult}_n(\bar{\mathcal{M}}, \dot{F}^{\bar{\mathcal{N}}})$ *agrees with* $\bar{\mathcal{N}}$ *below* $\mathrm{lh}(\dot{F}^{\bar{\mathcal{N}}})$, *while* $\mathrm{Ult}_n(\mathcal{M}, \dot{F}^{\mathcal{N}})$ *agrees with* $\mathcal{N}$ *below* $\mathrm{lh}(\dot{F}^{\mathcal{N}})$.
(c) $\sigma \restriction \mathrm{lh}(\dot{F}^{\bar{\mathcal{N}}}) + 1 = \psi \restriction \mathrm{lh}(\dot{F}^{\bar{\mathcal{N}}}) + 1$.
(d) *The diagram*

$$\begin{array}{ccc} \bar{\mathcal{M}} & \xrightarrow{\pi} & \mathcal{M} \\ \downarrow{\scriptstyle i} & & \downarrow{\scriptstyle j} \\ \mathrm{Ult}_n(\bar{\mathcal{M}}, \dot{F}^{\bar{\mathcal{N}}}) & \xrightarrow{\sigma} & \mathrm{Ult}_n(\mathcal{M}, \dot{F}^{\mathcal{N}}) \end{array}$$

commutes, where i *and* j *are the canonical* n*-embeddings.*

Remark. We want to allow the possibility $(\bar{\kappa}^+)^{\bar{\mathcal{M}}} = \mathrm{OR}^{\bar{\mathcal{M}}}$. In this case, we make our standard convention: $\pi(\mathrm{OR}^{\bar{\mathcal{M}}}) = \mathrm{OR}^{\mathcal{M}}$. We allow $\mathrm{lh}(\dot{F}^{\bar{\mathcal{N}}}) = \mathrm{OR}^{\bar{\mathcal{N}}}$ as well, and make a similar convention in (c) of the conclusion.

PROOF. The map σ is defined by

$$\sigma\left([a, f]^{\bar{\mathcal{M}}}_{\dot{F}\bar{\mathcal{N}}}\right) = [\psi(a), \pi(f)]^{\mathcal{M}}_{\dot{F}\mathcal{N}} \qquad \text{if } n = 0, \text{ and}$$

$$\sigma\left([a, f_{\tau,q}]^{\bar{\mathcal{M}}}_{\dot{F}\bar{\mathcal{N}}}\right) = [\psi(a), f_{\tau,\pi(q)}]^{\mathcal{M}}_{\dot{F}\mathcal{N}} \qquad \text{if } n > 0.$$

If X is the set used to show that π is a weak n-embedding then the set $i''X$ will show that σ is a weak n-embedding. It is straightforward to verify that this works. □

DEFINITION. If $\mathcal{T}$ and $\mathcal{U}$ are iteration trees then we say that $\vec{\pi} = \langle \pi_\alpha : \alpha < \mathrm{lh}(\mathcal{T})\rangle$ is a weak n-embedding from $\mathcal{T}$ to $\mathcal{U}$ if the following 6 conditions are satisfied.

(1) $T^{\mathcal{T}} = T^{\mathcal{U}}$, $\deg^{\mathcal{T}} = \deg^{\mathcal{U}}$ and $D^{\mathcal{T}} = D^{\mathcal{U}}$.
(2) $\pi_0\colon \mathcal{M}_0 \to \mathcal{N}_0$ is a weak n-embedding.
(3) For each ordinal α with $0 \leq \alpha < \mathrm{lh}\,\mathcal{T}$ there is a set Y such that $\pi_\alpha\colon \mathcal{M}_\alpha \to \mathcal{N}_\alpha$ is a $(\deg^{\mathcal{T}}(\alpha), Y)$-embedding, where $\mathcal{M}_\alpha$ and $\mathcal{N}_\alpha$ are the αth models of $\mathcal{T}$ and $\mathcal{U}$ respectively.
(4) $\pi_\alpha \restriction \mathrm{lh}\,E_\alpha + 1 = \pi_\delta \restriction \mathrm{lh}\,E_\alpha + 1$ whenever $\alpha < \delta < \theta'$.
(5) $\pi_\gamma \circ i^{\mathcal{T}}_{\alpha\gamma} = i^{\mathcal{U}}_{\alpha\gamma} \circ \pi_\alpha$ whenever $\alpha T \gamma$ and $(\alpha, \gamma]_T \cap D = \varnothing$.

DEFINITION. We say that $\vec{\pi}$ is a *tree embedding* if it is a weak n-embedding for some $n \leq \omega$ such that $\mathcal{T}$ is n-bounded if $n < \omega$.

Lemma. *Suppose that*

$$\mathcal{T} = \langle T, \deg, D, \langle E_\alpha, \mathcal{M}^*_{\alpha+1} \mid \alpha + 1 < \theta\rangle\rangle$$

is a n-maximal, n-bounded iteration tree on $\mathcal{M}$, where $n \leq \omega$, and that $\pi: \mathcal{M} \to \mathcal{N}$ is an weak n-embedding, where $\mathcal{N}$ is a n-iterable premouse. Then there is a tree $\pi\mathcal{T}$ on $\mathcal{N}$ and a tree embedding $\vec{\pi}: \mathcal{T} \to \pi\mathcal{T}$ such that $\pi_0 = \sigma$.

Proof. We define $\pi\mathcal{T} \restriction \alpha+1$ and π_α by recursion on $\alpha < \theta$. For $\beta = 0$ we have $\mathcal{N}_0 = \mathcal{N}$ and $\pi_0 = \pi$. Now suppose we have defined $\pi\mathcal{T} \restriction \beta$, together with sets Y_α such that π_α is a $(\deg^{\mathcal{T}}(\alpha), Y_\alpha)$-embedding for each ordinal $\alpha < \beta$.

If $\beta > 0$ is a limit ordinal then we set

$$\mathcal{N}_\beta = \operatorname{dir\,lim}\{\, \mathcal{N}_\alpha : \alpha T \beta \text{ and } D \cap [\alpha, \beta]_T = \varnothing \,\},$$

where the direct limit is taken along the maps $j_{\alpha\gamma}$, and we define π_β by setting $\pi_\beta(i_{\alpha\beta}(x)) = j_{\alpha,\beta}(\pi_\alpha(x))$ for $\alpha T \beta$ such that $[\alpha, \beta)_T \cap D = \varnothing$. Finally we set $Y_\alpha = i^{\mathcal{T}}_{\beta,\alpha}{}''Y_\beta$ for any $\beta T \alpha$ large enough that $i^{\mathcal{T}}_{\beta,\alpha}$ is defined.

For successor ordinals $\beta = \delta+1$, let $E_\delta = \dot{F}^{\mathcal{P}}$, where $\mathcal{P} = \mathcal{J}^{\mathcal{M}_\delta}_\eta$. Set $\mathcal{Q} = \mathcal{J}^{\mathcal{N}_\delta}_{\pi_\delta(\eta)}$, (with the usual convention if $\eta = \operatorname{dom} \pi_\delta$) and $F_\delta = \dot{F}^{\mathcal{Q}}$. Let T-Pred$(\delta+1) = \alpha$, let $\mathcal{M}^*_{\delta+1} = \mathcal{J}^{\mathcal{M}_\alpha}_\gamma$, and set $\mathcal{N}^*_{\delta+1} = \mathcal{J}^{\mathcal{N}_\alpha}_{\pi_\alpha(\gamma)}$, again with the usual convention if $\gamma = \operatorname{dom} \pi_\alpha$.

We will use the shift lemma, to define $\pi_{\delta+1}$. Let σ be the natural embedding of $\mathcal{M}^*_{\delta+1}$ into $\mathcal{N}^*_{\delta+1}$. Let $\bar{\kappa} = \operatorname{crit} E_\gamma$. Then $(\bar{\kappa}^+)^{\mathcal{M}^*_{\delta+1}} \leq \operatorname{lh} E_\alpha$ (possibly with $(\bar{\kappa}^+)^{\mathcal{M}^*_{\delta+1}} = \mathrm{OR}^{\mathcal{M}^*_{\delta+1}}$), so σ and π_γ agree up to and at $(\bar{\kappa}^+)^{\mathcal{M}^*_{\delta+1}}$. Thus we can apply the shift lemma to get $\pi_{\delta+1}: \mathcal{M}_{\delta+1} \to \mathcal{N}_{\delta+1}$ satisfying our inductive hypotheses on commutativity and agreement. If $\mathcal{M}^*_{\delta+1} = \mathcal{M}_\alpha$ and $\deg^{\mathcal{T}}(\delta+1) = \deg^{\mathcal{T}}(\alpha)$ then set $Y_{\delta+1} = i^{\mathcal{T}}_{\alpha,\delta+1}{}''Y_\alpha$. Otherwise take $Y_{\delta+1} = i^{*\mathcal{T}}_\beta{}''\mathcal{M}^*_\beta$. To see $\pi_{\delta+1}$ is a $\deg(\delta+1, Y_{\delta+1})$-embedding when T-Pred$(\delta+1) = 0$, use n-boundedness, and for T-Pred$(\delta+1) > 0$. Note that σ is fully elementary if $\mathcal{M}^*_{\delta+1} \neq \mathcal{M}_\alpha$, and that $\deg^{\mathcal{T}}(\delta+1) < \deg^{\mathcal{T}}(\alpha)$ if the degrees are not equal.

This finishes the recursive definition of $\pi\mathcal{T}$, and it only remains to verify that each $\mathcal{N}_\beta$ is well founded. Suppose that it is not. Since $\mathcal{N}$ is n-iterable, it follows that there is another branch b in T, cofinal in β, such that if $\mathcal{N}_b$ is the limit along the branch b in $\mathcal{U}$ then $\mathcal{N}_b$ is well founded. This is impossible since there is an embedding $\pi_b: \mathcal{M}_b \to \mathcal{N}_b$, where $\mathcal{M}_b$ is the limit in $\mathcal{T}$ along the branch b, and $\mathcal{M}_b$ is ill founded since $\mathcal{T}$ is simple and $\mathcal{M}_\beta$ is well founded. □

The Dodd-Jensen Lemma. We are now ready to prove the Dodd-Jensen lemma on the minimality of iteration maps. This is a powerful tool which will be crucial in what follows. We shall call it simply the Dodd-Jensen lemma, though without meaning to suggest that this is the most important of the lemmas which they have proved. Our proof is just the obvious generalization of the original proof of Dodd and Jensen.

Lemma 5.3 (Dodd-Jensen Lemma). *Let $\mathcal{T} = \langle T, deg, D, \langle E_\alpha, \mathcal{M}^*_{\alpha+1} \mid \alpha+1 < \theta+1\rangle\rangle$ be an n-bounded, simple iteration tree of length $\theta+1$ on a n-iterable* *premouse* $\mathcal{M}_0$. *Suppose*

$$\sigma : \mathcal{M}_0 \to Q$$

is an weak n-embedding, where $n \leq \omega$ *and* Q *is an initial segment of* $\mathcal{M}_\theta$. *Then*

(1) $Q = \mathcal{M}_\theta$.

Moreover, if there is an ordinal $\gamma \in [0,\theta]_T$ *such that* $\deg(\gamma') \geq n$ *whenever* $\gamma' \in [\gamma,\theta]_T$ *then the following two clauses hold in addition:*

(2) $D \cap [0,\theta]_T = \emptyset$, *so that* $\deg(\gamma) = n$ *for all* $\gamma \in [0,\theta]_T$,

(3) $i_{0,\theta}(\eta) \leq \sigma(\eta)$, *for all* $\eta \in OR \cap \mathcal{M}_0$.

Remark. Notice that the additional precondition for clauses (2) and (3) is equivalent to the condition that $\mathcal{M}_\theta$ is n-sound. This equivalence will be used in many of our applications: We will know from the construction of σ that Q is n-sound, so that clause (1) implies that $\mathcal{M}_\theta = Q$ and hence $\mathcal{M}_\theta$ is n-sound so that that clauses (2) and (3) of the lemma must are valid as well.

PROOF. We will define a sequence $\langle \mathcal{T}_i : i < \omega \rangle$ of iteration trees as in clause (b) of the definition of k-iterable, together with maps $\sigma_i : \mathcal{M}^i_0 \to \mathcal{M}^i_\theta$ where $\mathcal{M}^i_\gamma$ is the γth model of $\mathcal{T}_i$. For each integer i the pair $(\mathcal{T}_i, \sigma_i)$ will satisfy the same conditions as the pair $(\mathcal{T}_0, \sigma_0) = (\mathcal{T}, \sigma)$, and it will follow that any failure of the lemma will imply that $\langle \mathcal{T}_i : i < \omega \rangle$ violates condition (b) of the definition of k-iterable.

We first give the definition under the assumption $Q = \mathcal{M}_\theta$. We will then modify the definition slightly to prove that $Q = \mathcal{M}_\theta$.

We have $\mathcal{T}_0 = \mathcal{T}$ and $\sigma_0 = \sigma$. Now suppose we are given a simple, n-bounded tree $\mathcal{T}_i$ on the n-iterable model M^i_0, together with a (n, X_i)-embedding $\sigma_i : \mathcal{M}^i_0 \to \mathcal{M}^i_\theta$. Let

$$\mathcal{T}_{i+1} = \sigma_i \mathcal{T}_i .$$

Since $\mathcal{M}^i_0$ is n-iterable and $\mathcal{T}_i$ is simple, $\mathcal{M}^i_\theta = \mathcal{M}^{i+1}_0$ is n-iterable. Thus $\mathcal{T}_{i+1}$ has length $\theta + 1$ and is simple and n-bounded. Let $\vec{\pi}^i : \mathcal{T}_i \to \sigma_i \mathcal{T}_i = \mathcal{T}_{i+1}$ be the tree embedding given by the copying procedure, and set

$$\sigma_{i+1} = \pi^i_\theta : \mathcal{M}^i_\theta \to \mathcal{M}^{i+1}_\theta .$$

Since $\deg(\gamma + 1) \geq n$ for all sufficiently large $\gamma + 1 \in [0,\theta]_T$, σ_{i+1} is a (n, X_{i+1})-embedding, where X_{i+1} is given by the copying procedure. Thus we are ready for the next stage of the construction.

This completes the definition of the $\mathcal{T}_i$'s and σ_i's. We must have $D \cap [0,\theta]^{\mathcal{T}} \neq \emptyset$, since otherwise $D_i \cap [0,\theta]_{\mathcal{T}_i} \neq \emptyset$ for all $i < \omega$, contradicting clause (b) of the definition of k-iterable. Thus there are canonical n-embeddings

$$\tau_i : \mathcal{M}^i_0 \to \mathcal{M}^i_\theta$$

given by composing the embeddings along the branch $[0,\theta]_T$ of $\mathcal{T}_i$. We have the commutative diagram

$$\begin{array}{ccccccc}
\mathcal{M}_0^0 & \xrightarrow{\tau_0} & \mathcal{M}_\theta^0 = \mathcal{M}_0^1 & \xrightarrow{\tau_1} & \mathcal{M}_\theta^1 = \mathcal{M}_0^2 & \xrightarrow{\tau_2} & \cdots \\
 & & \uparrow{\scriptstyle \sigma_0} & & \uparrow{\scriptstyle \sigma_1} & & \\
 & & \mathcal{M}_0^0 & \xrightarrow{\tau_0} & \mathcal{M}_0^1 & \xrightarrow{\tau_1} & \cdots
\end{array}$$

Suppose toward a contradiction that $i_{0,\theta}(\eta_0) = \tau_0(\eta_0) > \sigma(\eta_0)$. Set $\eta_{i+1} = \sigma_i(\eta_i)$. It is routine to check that $\tau_i(\eta_i) > \sigma_i(\eta_i) = \eta_{i+1}$ for all i and it follows that $\operatorname{dir\,lim}(\mathcal{M}_i : i < \omega)$ is not well founded, contradicting clause (b) in the definition of n-iterability of $\mathcal{M}_0^\theta$.

To show $Q = \mathcal{M}_\theta$ we proceed essentially as above. If $Q \neq \mathcal{M}_\theta$ we will have $\sigma_i : \mathcal{M}_i^0 \to Q_i$, with $Q_0 = Q$ and Q_i a proper initial segment of $\mathcal{M}_i^\theta$. In this case $\sigma_i \mathcal{T}_i$ is a tree on Q_i rather than $\mathcal{M}_\theta^i$, but it can be modified slightly to make it a tree on $\mathcal{M}_\theta$ which immediately drops to Q_i at all $\mathcal{T}_0$ successors of 0. That is, $\mathcal{T}_{i+1}$ is the same as $\sigma_i \mathcal{T}_i$ except that we put $\gamma+1$ into D_1 whenever $T\text{-pred}(\gamma+1) = 0$, and we set $(\mathcal{M}_{\gamma+1}^1)^* = Q_i$ or the appropriate initial segment thereof. With this modification the construction works as before, giving a sequence of trees $(\mathcal{T}_i : i < \omega)$ such that $D_i \cap [0, \theta]_{T_i} \neq \varnothing$ for every $i > 0$ and thus contradicting clause (b) of the definition of n-iterability. Notice that in this case we don't need the hypothesis that $\deg(\gamma + 1) \geq n$ for all sufficiently large $\gamma + 1 \in [0, \theta]$, since for example it is not σ_i but $\sigma_i \restriction Q_i$ which will be used to produce $\mathcal{T}_{i+1}$, and $\sigma_i \restriction Q_i$ is fully elementary. □

§6. Uniqueness of Wellfounded Branches

We shall show that, roughly speaking, all iteration trees which are important for the comparison of 1-small mice are simple.

Let $\mathcal{T} = \langle T, \deg, D, \langle E_\alpha, \mathcal{M}^*_{\alpha+1} \mid \alpha + 1 < \theta\rangle\rangle$ be an iteration tree of length θ. We set

$$\vec{E}(\mathcal{T}) = \bigcup_{\alpha<\theta} (\dot{E}^{\mathcal{M}_\alpha} \restriction \operatorname{lh} E_\alpha)$$

$$\delta(\mathcal{T}) = \bigcup_{\alpha<\theta} \operatorname{lh} E_\alpha$$

By 5.1, $\dot{E}^{\mathcal{M}_\alpha} \restriction \operatorname{lh} E_\alpha = \dot{E}^{\mathcal{M}_\beta} \restriction \operatorname{lh} E_\alpha$ for all $\beta > \alpha$, so that $\vec{E}(\mathcal{T})$ is a good extender sequence with domain included in $\delta(\mathcal{T})$. Notice that if b is a cofinal wellfounded branch of $\mathcal{T}$, then $\vec{E}(\mathcal{T}) = \dot{E}^{\mathcal{M}_b} \restriction \delta(\mathcal{T})$.

Theorem 6.1 (Uniqueness Theorem). *Let $\mathcal{T}$ be an iteration tree of limit length θ, and b and c be distinct cofinal wellfounded branches of $\mathcal{T}$. Let $\alpha = OR^{\mathcal{M}_b} \cap OR^{\mathcal{M}_c}$, so that $\alpha \geq \delta(\mathcal{T})$, and suppose that $\alpha > \delta(\mathcal{T})$. Then*

$$J_\alpha^{\vec{E}(\mathcal{T})} \models \delta(\mathcal{T}) \ \textit{ is Woodin}.$$

Proof. Just as in [MS]. Here is a slightly cleaner presentation of that argument, adapted to our context.

Let $\delta = \delta(\mathcal{T})$, $\vec{E} = \vec{E}(\mathcal{T})$, and let $f : \delta \to \delta$ with $f \in J_\alpha^{\vec{E}}$. Let $\beta < \theta$ be large enough that

$$D \cap (b \cup c) \subseteq \beta$$

and

$$b \cap \beta \neq c \cap \beta$$

and

$$\gamma \in b - \beta \Rightarrow f, \vec{E}, \delta \in \operatorname{ran} i_{\gamma b},$$
$$\gamma \in c - \beta \Rightarrow f, \vec{E}, \delta \in \operatorname{ran} i_{\gamma c},$$

and $\alpha \in \operatorname{ran} i_{\gamma b}$ if $\alpha \neq \mathrm{OR}^{\mathcal{M}_b}$, and $\alpha \in \operatorname{ran} i_{jc}$ if $\alpha \neq \mathrm{OR}^{\mathcal{M}_c}$.

Claim 1. If $\gamma \in b - \beta$ and $\eta \in c - \beta$, then

$$(\operatorname{ran} i_{\gamma b} \cap \operatorname{ran} i_{\eta c} \cap J_\alpha^{\vec{E}}) \prec_{\Sigma_1} J_\alpha^{\vec{E}}.$$

Proof. Straightforward. The restriction to Σ_1 is due to the limited elementarity of the maps $i_{\gamma b}, i_{\eta c}$.

CLAIM 2. Let $\gamma+1 \in b$ with T-pred$(\gamma+1)=\xi \geq \beta$, and let η be a member of c such that $\beta < c < \gamma+1$ such that if $c < \xi$ then η is the largest member of c such that $\eta < \gamma+1$. Then

$$\operatorname{ran} i_{\xi b} \cap \operatorname{ran} i_{\eta c} \cap \delta = \inf\{\operatorname{crit} i_{\xi b},\ \operatorname{crit} i_{\eta c}\}\,.$$

PROOF. $\supseteq$ is obvious. Let us define

$$\begin{aligned}\gamma_0 &= \gamma+1\\ \eta_n &= \text{least ordinal in } c-\gamma_n\\ \gamma_{n+1} &= \text{least ordinal in } b-\eta_n\end{aligned}$$

for all $n<\omega$. The γ_n's and η_n's are all successor ordinals. Also we have $\sup_{n<\omega}\gamma_n = \sup_{n<\omega}\eta_n$, so the common sup is θ. Notice also that T-pred$(\eta_n) < \gamma_n$ and T-pred$(\gamma_{n+1}) < \eta_n$ by the minimality of our choices. Also T-pred$(\eta_0)=\eta$ (unless $\eta \geq \xi$ in which case this may fail), and T-pred$(\gamma_0)=\xi$.

Now suppose $\mu \in \operatorname{ran} i_{\xi b} \cap \operatorname{ran} i_{\eta c} \cap \delta$. As $\mu < \delta$, we have an $n<\omega$ such that

$$\mu < \operatorname{lh} E_{\gamma_{n+1}-1}\,.$$

Since $\mu \in \operatorname{ran} i_{\xi b}$ and $\xi T \gamma_{n+1}$,

$$\mu < \operatorname{crit} E_{\gamma_{n+1}}\,.$$

By clauses (3) and (4) on iteration trees,

$$\mu < \operatorname{lh} E_{T\text{-pred}(\gamma_{n+1})} \leq \operatorname{lh} E_{\eta_n-1}\,.$$

Since $\mu \in \operatorname{ran} i_{\eta c}$ and $\eta T \eta_n$,

$$\mu < \operatorname{crit} E_{\eta_n-1}\,.$$

By clauses (3) and (4) on iteration trees

$$\mu < \operatorname{lh} E_{T\text{-pred}(\eta_n)} \leq \operatorname{lh} E_{\gamma_n-1}\,.$$

So we may repeat the cycle until we get $\mu < \operatorname{lh} E_{\gamma_0-1}$. Then applying the argument again we get

$$\mu < \operatorname{crit} E_{\gamma_0-1} < \operatorname{lh} E_\xi\,.$$

So if $\nu+1 \in b-(\xi+1)$ or $\nu+1 \in c-(\eta+1)$ then $\nu \geq \xi$ (under either hypothesis on η) so that $\mu < \operatorname{lh} E_\nu$, so $\mu < \operatorname{crit} E_\nu$. Thus $\mu < \operatorname{crit} i_{\eta c}$ and $\mu < \operatorname{crit} i_{\xi b}$.

CLAIM 3. Claim 2 holds with the roles of b and c reversed.

PROOF. The proof is the same as that of claim 2.

Now fix $\beta' > \beta$ such that $b \cap (\beta' - \beta) \neq \emptyset$ and $c \cap (\beta' - \beta) \neq \emptyset$. Let

$$\kappa = \text{ least } \nu \text{ such that } \nu = \text{crit } E_\gamma \text{ for some } \gamma + 1 \in (b \cup c) - \beta' .$$

Let γ be largest such that $\kappa = \text{crit } E_\gamma$ and $\gamma + 1 \in (b \cup c) - \beta'$, and suppose without loss of generality that $\gamma + 1 \in b$. Let η be the largest element of c which is $< \gamma + 1$. Notice crit $i_{\eta c} =$ crit E_ν for some $\nu + 1 \in c$ such that $\gamma + 1 < \nu + 1$; thus crit $i_{\eta c} > \kappa$. So

$$\kappa = \text{ran } i_{\eta c} \cap \text{ran } i_{\xi b} \cap \delta$$

where $\xi = T\text{-pred}(\gamma + 1)$, and it follows by Claim 1 that κ is closed under f. Now let $\nu = \inf\{\text{crit } i_{\eta c}, \text{crit } i_{\gamma+1,b}\}$ Claim 3 implies that

$$\nu = \text{ran } i_{\eta c} \cap \text{ran } i_{\gamma+1,b} \cap \delta$$

so that ν is closed under f. Note also that $\kappa < \nu$.

We claim that $\nu < \rho_\gamma$. (Recall that ρ_γ is the sup of the generators for E_γ.) Let $\tau \in c$ and $T\text{-pred}(\tau) = \eta$. Then $\nu \leq \text{crit } i_{\eta c} \leq \text{crit } E_{\tau-1} < \rho_\eta$. So if $\eta = \gamma$ we're done. Otherwise $\eta < \gamma$, so lh E_η is a cardinal of $\mathcal{M}_\gamma$, and as lh $E_\eta <$ lh E_γ, lh $E_\eta \leq \rho_\gamma$. As $\nu < \rho_\eta$, $\nu < \rho_\gamma$.

Our initial segment condition on good extender sequences implies that $E_\gamma \restriction \nu$ is an initial segment of some extender F which is on the sequence of $\mathcal{M}_\gamma$ before E_γ. By coherence we see that F is one of the extenders on $\vec{E} = \vec{E}(\mathcal{T})$. So $E_\gamma \restriction \nu \in J_\alpha^{\vec{E}}$.

We leave it to the reader to check that ν is an inaccessible cardinal of $J_\alpha^{\vec{E}}$. By strong acceptability and the fact that F coheres with $\vec{E}$,

$$J_\alpha^{\vec{E}} \models \text{“}V_\nu \in \text{Ult}(V, E_\gamma \restriction \nu)\text{”} .$$

Finally, suppose $i_{\xi b}(\bar{f}) = f$. Then $\bar{f} \restriction \kappa = f \restriction \kappa$, and

$$i_{\xi,\gamma+1}(\bar{f}) \restriction \nu = f \restriction \nu$$

so

$$i_{\xi,\gamma+1}(f \restriction \kappa)(\kappa) < \nu .$$

But

$$i_{\xi,\gamma+1}(f \restriction \kappa) \restriction \nu = i_{E_\gamma \restriction \nu}(f \restriction \kappa) \restriction \nu$$

as computed in $J_\alpha^{\vec{E}}$. Thus $E_\gamma \restriction \nu$ witnesses that δ is Woodin with respect to f in $J_\alpha^{\vec{E}}$. □

For the purpose of comparison we are only interested in iteration trees in which each E_α is applied to the earliest model to which it can be.

DEFINITION 6.1.1. $\mathcal{T} = \langle T, \deg, D, \langle E_\alpha, \mathcal{M}^*_{\alpha+1} \mid \alpha+1 < \theta\rangle\rangle$ is *non-overlapping* iff whenever T-pred$(\gamma+1) = \beta$, then $\rho_\eta \leq \text{crit } E_\gamma$ for all $\eta < \beta$.

Here ρ_η is the sup of the generators for E_η, so that crit $E_\gamma < \rho_\beta$. Clearly, generators are not moved along the branches of a nonoverlapping tree, and in fact not moving generators is equivalent to being non-overlapping.

We want also to restrict ourselves to trees in which $\mathcal{M}^*_{\gamma+1}$ and $\deg(\gamma+1)$ are as large as possible, subject perhaps to an n-boundedness requirement.

DEFINITION 6.1.2. Let $\mathcal{T} = \langle T, \deg, D, \langle E_\alpha, \mathcal{M}^*_{\alpha+1} \mid \alpha+1 < \theta\rangle\rangle$ be an iteration tree, and $n \leq \omega$. We say $\mathcal{T}$ is *n-maximal* iff $\mathcal{T}$ is non-overlapping, and whenever T-pred$(\gamma+1) = \beta$, $E_\gamma = \dot{F}^{\mathcal{N}}$ where $\mathcal{N}$ is an initial segment of $\mathcal{M}_\gamma$, and $\kappa =$ crit E_γ, then

(a) $\mathcal{M}^*_{\gamma+1}$ is the longest initial segment $\mathcal{P}$ of $\mathcal{M}_\beta$ such that $P(\kappa) \cap |\mathcal{P}| = P(\kappa) \cap |\mathcal{N}|$, and

(b) if $D \cap [0, \gamma+1]_T = \varnothing$ then $\deg(\gamma+1)$ is the largest integer $k \leq n$ such that $\kappa < \rho_k^{\mathcal{M}^*_{\gamma+1}}$, and

(c) if $D \cap [0, \gamma+1]_T \neq \varnothing$, then $\deg(\gamma+1)$ is the largest $k \in \omega$ such that $\kappa < \rho_k^{\mathcal{M}^*_{\gamma+1}}$.

Notice that in (a) of the definition $\mathcal{P}$ is the longest initial segment Q of $\mathcal{M}_\beta$ such that

$$P(\kappa) \cap J^{\mathcal{M}_\beta}_{\text{lh } E_\beta} = P(\kappa) \cap Q\,.$$

Since $J^{\mathcal{M}_\beta}_{\text{lh } E_\beta} = J^{\mathcal{M}_\gamma}_{\text{lh } E_\beta}$ it follows that if $\beta \neq \gamma$ then $\mathcal{P}$ is the longest initial segment Q of $\mathcal{M}_\beta$ such that $P(\kappa) \cap Q = P(\kappa) \cap |\mathcal{M}_\gamma|$.

The iteration trees for which we have any practical use are all n-maximal for some $n \leq \omega$. One important elementary property of such trees is the following.

Lemma 6.1.5. *Let $\mathcal{T} = \langle T, \deg, D, \langle E_\alpha, \mathcal{M}^*_{\alpha+1} \mid \alpha+1 < \theta\rangle\rangle$ be an n-maximal iteration tree, where $n \leq \omega$; then for any $\alpha+1 < \theta$, E_α is close to $\mathcal{M}^*_{\alpha+1}$.*

PROOF. By induction on α. Let $\beta = T$-pred$(\alpha+1)$. We may assume $\beta < \alpha$; otherwise E_α is on the $\mathcal{M}_\beta$ sequence, and so by the restrictions on how far $\mathcal{M}^*_{\alpha+1}$ can drop in $\mathcal{M}_\beta$, on the $\mathcal{M}^*_{\alpha+1}$ sequence. Thus E_α is close indeed to $\mathcal{M}^*_{\alpha+1}$.

Let $a \subseteq \text{lh } E_\alpha$ be finite. We wish to verify the two conditions in closeness to $\mathcal{M}^*_{\alpha+1}$ for $(E_\alpha)_a$. We begin with the second.

Let $\kappa =$ crit E_α and $\tau = \text{lh } E_\beta$. As $\beta = T$-pred$(\alpha+1)$, $\kappa < \tau$, and as τ is a cardinal of $\mathcal{M}_\alpha$, $(\kappa^+)^{\mathcal{M}_\alpha} \leq \tau$. Let $A \subseteq P([\kappa]^{\text{card } a})$, $A \in |\mathcal{M}^*_{\alpha+1}|$, be such that $\mathcal{M}^*_{\alpha+1} \models \text{card}(A) \leq \kappa$. We want to see that $(E_\alpha)_a \cap A \in |\mathcal{M}^*_{\alpha+1}|$. Now $P(\kappa) \cap |\mathcal{M}_\alpha| = P(\kappa) \cap |\mathcal{M}^*_{\alpha+1}|$, so A has cardinality $\leq \kappa$ in $\mathcal{M}_\alpha$. But then $(E_\alpha)_a \cap A$ is in $\mathcal{M}_\alpha$ and has cardinality $\leq \kappa$ there, by weak amenability. But then $(E_\alpha)_a \cap A \in |\mathcal{M}^*_{\alpha+1}|$, as desired.

It remains to show $(E_\alpha)_a$ is Σ_1 over $\mathcal{M}^*_{\alpha+1}$. The following claim is useful; notice that $\mathcal{J}_\tau^{\mathcal{M}_\beta}$ is an initial segment of $\mathcal{M}^*_{\alpha+1}$.

CLAIM 1. If $A \subseteq \tau$ and $A \in |\mathcal{M}_\gamma|$ for some $\gamma > \beta$, then A is Σ_1 over $\mathcal{J}_\tau^{\mathcal{M}_\beta}$.

PROOF. By 5.1, $A \in |\mathcal{M}_{\beta+1}|$. Let $A = [a, f]^{\mathcal{Q}}_{E_\beta}$, where $\mathcal{Q} = \mathcal{M}^*_{\alpha+1}$. Since $A \subseteq \tau$, we can take f to map $[\mu]^{\operatorname{card} a}$ into $J^{\mathcal{Q}}_\mu$, where $\mu = \operatorname{crit} E_\beta$. We can therefore assume $f \in |\mathcal{Q}|$, as $\mu < \rho^{\mathcal{Q}}_m$ where $\mathcal{M}_{\beta+1} = \operatorname{Ult}_m(\mathcal{Q}, E_\beta)$. But also, $\mathcal{M}_\beta$ agrees with $\mathcal{Q}$ below τ, and $f \in J_\tau^{\mathcal{M}_\beta} = \mathcal{P}$. Moreover, $A = [a, f]^{\mathcal{Q}}_{E_\beta} = [a, f]^{\mathcal{P}}_{E_\beta}$. It is easy, then, to define A in a Σ_1 way over $\mathcal{P}$ from the parameters a and f. □

It follows that if $(E_\alpha)_a \in |\mathcal{M}_\alpha|$, then since $(E_\alpha)_a$ is coded by a subset of τ, $(E_\alpha)_a$ is Σ_1 over $\mathcal{J}_\tau^{\mathcal{M}_\beta}$, hence Σ_1 over $\mathcal{M}^*_{\alpha+1}$, as required. Thus we may assume that $(E_\alpha)_a \notin |\mathcal{M}_\alpha|$, and hence E_α is on the $\mathcal{M}_\alpha$ sequence, $\mathcal{M}_\alpha$ is active and $E_\alpha = \dot{F}^{\mathcal{M}_\alpha}$.

CLAIM 2. Let $\gamma \in [0, \alpha]_T$ be such that $\gamma \geq \beta$ and $D \cap (\gamma, \alpha]_T = \emptyset$. Then $\operatorname{crit}(i_{\gamma\alpha}) > \kappa$, and $(E_\alpha)_a$ is Σ_1 over $\mathcal{M}_\gamma$. If, in addition, $\gamma > \beta$ and γ is a successor ordinal, then $\operatorname{crit}(i_{\gamma,\alpha} \circ i^*_\gamma) > \kappa$ and $(E_\alpha)_a$ is Σ_1 over $\mathcal{M}^*_\gamma$.

PROOF. Since $\kappa = \operatorname{crit} E_\alpha$ and $E_\alpha = \dot{F}^{\mathcal{M}_\alpha}$, $\kappa \in \operatorname{ran} i_{\gamma\alpha}$. On the other hand, every extender used in $i_{\gamma\alpha}$ has length at least $\operatorname{lh} E_\beta$, since $\gamma \geq \beta$. It follows that $\kappa < \operatorname{crit}(i_{\gamma\alpha})$.

By our induction hypothesis, E_η is close to $\mathcal{M}^*_{\eta+1}$ for all $\eta < \alpha$. Thus the preservation facts recorded in 4.5, 4.6, and 4.7 hold for the embeddings of $\mathcal{T} \restriction (\alpha+1)$. Now $\rho_1^{\mathcal{M}_\alpha} \leq \tau = (\kappa^+)^{\mathcal{M}_\alpha}$ since $(E_\alpha)_a \notin |\mathcal{M}_\alpha|$, and $\tau \leq \operatorname{crit} i_{\gamma\alpha}$, so $\deg(\eta) = 0$ for all $\eta \in (\gamma, \alpha]_T$. The proofs of 4.5 and 4.7 (see especially 4.5) show that every $\Sigma_1^{\mathcal{M}_\alpha}$ subset of $\operatorname{crit}(i_{\gamma\alpha})$ is $\Sigma_1^{\mathcal{M}_\gamma}$. Thus $(E_\alpha)_a$ is $\Sigma_1^{\mathcal{M}_\gamma}$, as desired.

Suppose finally that $\gamma > \beta$ and γ is a successor ordinal. The extenders used in $i_{\gamma\alpha} \circ i^*_\gamma$ are just those used in $i_{\gamma\alpha}$ together with $E_{\gamma-1}$. Since $\gamma - 1 \geq \beta$, all these have length at least $\operatorname{lh} E_\beta$, hence $> \kappa$. The argument of the previous paragraph now shows $\operatorname{crit}(i_{\gamma\alpha} \circ i^*_\gamma) > \kappa$ and $(E_\alpha)_a$ is Σ_1 over $\mathcal{M}^*_\gamma$. □

Now let $\eta \in [0, \alpha]_T$ be least such that $\beta \leq \eta$. Suppose first that $D \cap (\eta, \alpha]_T \neq \emptyset$. Let γ be largest in $D \cap (\eta, \alpha]_T$, and $\xi = T\text{-pred}(\gamma)$. Since $\gamma > \beta$, Claim 2 implies that $(E_\alpha)_a$ is Σ_1 over $\mathcal{M}^*_\gamma$. Since $\gamma \in D$, $\mathcal{M}^*_\gamma \in |\mathcal{M}_\xi|$, so $(E_\alpha)_a \in |\mathcal{M}_\xi|$. Since $\xi \geq \beta$, Claim 1 implies that $(E_\alpha)_a$ is Σ_1 over $\mathcal{M}^*_{\alpha+1}$, as desired.

So we may assume $D \cap (\eta, \alpha]_T = \emptyset$. We claim that $\eta = \beta$. For if $\eta > \beta$, then the leastness of η implies that η is not a limit, so let $\delta = T\text{-pred}(\eta)$. Since η is least, $\delta < \beta$. By Claim 2 with $\gamma = \eta$, $\operatorname{crit}(i^*_\eta) = \operatorname{crit}(E_{\eta-1}) > \kappa$. But $\operatorname{crit}(E_{\eta-1}) < \rho_\delta$, so $\kappa < \rho_\delta$. But the rules for non-overlapping trees then require that $T\text{-pred}(\alpha+1) \leq \delta$, a contradiction.

So $\eta = \beta$. Also, by Claim 2, $\operatorname{crit} i_{\beta\alpha} > \kappa$, and $(E_\alpha)_a$ is Σ_1 over $\mathcal{M}_\beta$. But then

$P(\kappa) \cap |\mathcal{M}_\beta| = P(\kappa) \cap |\mathcal{M}_\alpha|$, and since $\mathcal{T}$ is n-maximal, $\mathcal{M}_\beta = \mathcal{M}^*_{\alpha+1}$. Thus $(E_\alpha)_a$ is Σ_1 over $\mathcal{M}^*_{\alpha+1}$, as desired. □

Lemma 6.1.5 has the important consequence that the preservation facts listed in 4.5, 4.6, and 4.7 apply to the embeddings along the branches of an n-maximal tree. We shall use this repeatedly and without explicit mention in the future.

The following is a crucial strengthening of the uniqueness theorem (6.1). It will imply that only simple iteration trees arise in our proof that 1-small, k-iterable premice are k-solid for all k. This is important because our proof of that fact uses heavily the Dodd-Jensen lemma, which requires a simplicity hypothesis.

If $\mathcal{M}$ is a ppm, an "extender from the $\mathcal{M}$-sequence" is an extender E such that $E = \dot{F}^{\mathcal{M}}$ or E is on the sequence $\dot{E}^{\mathcal{M}}$.

Theorem 6.2 (Strong uniqueness). *Let $\mathcal{M}$ be an n-sound, 1-small n-iterable premouse and $\rho^{\mathcal{M}}_{n+1} \leq \operatorname{lh} E$ for some extender E from the $\mathcal{M}$-sequence and some integer n. Let $\mathcal{T}$ be an n-maximal iteration tree on $\mathcal{M}$. Then $\mathcal{T}$ is simple.*

PROOF. Assume toward a contradiction that b and c are distinct cofinal well-founded branches of $\mathcal{T}$ with $\mathrm{OR}^{\mathcal{M}_b} \leq \mathrm{OR}^{\mathcal{M}_c}$. Let $\delta = \delta(\mathcal{T})$.

CLAIM 1. $\operatorname{lh} F < \delta$ for all extenders F from the $\mathcal{M}_b$ sequence.

PROOF. Let F be the first extender on the $\mathcal{M}_b$ sequence such that $\operatorname{lh} F \geq \delta$. Notice δ is a limit of $\mathcal{M}_b$ cardinals, as crit i_{ab} is an $\mathcal{M}_b$ cardinal whenever i_{ab} is defined. Thus $\operatorname{lh} F > \delta$, as $\exists \nu < \operatorname{lh} F \forall \gamma < \operatorname{lh} F\, (\mathcal{M}_b \models \operatorname{card} \gamma \leq \nu)$. Let $\gamma = \operatorname{lh} F$. By Theorem 6.1,

$$J^{\vec{E}(\mathcal{T})}_\gamma \models \delta \text{ is Woodin}$$

so

$$\mathcal{J}^{\mathcal{M}_b}_\gamma = (J^{\vec{E}(\mathcal{T})}_\gamma, \in, \vec{E}(\mathcal{T}), \tilde{F}) \models \delta \text{ is Woodin}.$$

Now let $\mathcal{N} = \mathrm{Ult}_0(\mathcal{J}^{\mathcal{M}_b}_\gamma, F)$. As F is a pre-extender over $\mathcal{J}^{\mathcal{M}_b}_\gamma$, $\gamma \in \mathrm{wfp}(\mathcal{N})$. By coherence and strong acceptability and the fact that γ is a cardinal of $\mathcal{N}$,

$$\mathcal{N} \models \delta \quad \text{is Woodin.}$$

But then $\mathcal{N}$ is not 1-small, so that $\mathcal{M}_b$ is not 1-small and hence $\mathcal{M}$ is not 1-small, which is a contradiction. □

CLAIM 2. $\mathcal{M}_b$ is an initial segment of $\mathcal{M}_c$.

PROOF. Otherwise $\mathcal{M}_c$ is not 1-small. For let F be the first extender from the $\mathcal{M}_c$ sequence with $\operatorname{lh} F \geq \delta$; if none exists Claim 2 is obvious from Lemma 5.1. So $\operatorname{lh} F > \delta$ as in Claim 1. If $\mathcal{M}_b$ is not an initial segment of $\mathcal{M}_c$, $\operatorname{lh} F \leq \mathrm{OR}^{\mathcal{M}_b}$. But now we can show $\mathcal{M}_c$ is not 1-small as in Claim 1. □

CLAIM 3. If $\mathrm{OR}^{\mathcal{M}_b} < \mathrm{OR}^{\mathcal{M}_c}$, then there is no dropping of any kind along b; that is, $D^{\mathcal{T}} \cap b = \emptyset$ and $\deg^{\mathcal{T}}(\alpha+1) = n$ for all $\alpha + 1 \in b$.

PROOF. If $\mathrm{OR}^{\mathcal{M}_b} < \mathrm{OR}^{\mathcal{M}_c}$, then $\mathcal{M}_b$ is a proper initial segment of $\mathcal{M}_c$, and hence $\mathcal{M}_b$ is ω-sound since $\mathcal{M}_c$ is a premouse. But now suppose the last drop of any kind along b occurs at $\alpha+1$. Then $\alpha+1 \in b$, and $k = \deg(\alpha+1) = \deg(\gamma)$ for all $\gamma \in b - (\alpha+1)$. Also, $\mathcal{M}^*_{\alpha+1}$ is $k+1$ sound and $\mathrm{crit}(i_{\alpha+1,b} \circ i^*_{\alpha+1}) = \mathrm{crit}(i^*_{\alpha+1}) \geq \rho_{k+1}^{\mathcal{M}^*_{\alpha+1}}$. From Lemma 4.7 it follows that $\mathcal{M}_b$ is not $k+1$-sound, a contradiction. □

CLAIM 4. If $\mathrm{OR}^{\mathcal{M}_b} = \mathrm{OR}^{\mathcal{M}_c}$, then on one of b and c there's no dropping of any kind.

PROOF. Suppose the last drop along b occurs at $\eta+1$, and the last drop along c at $\gamma+1$. Since $\mathcal{M}_b = \mathcal{M}_c$, $\deg(\eta+1) = \deg(\gamma+1) = k$, where $k < \omega$ is least such that $\mathcal{M}_b = \mathcal{M}_c$ is not $k+1$-sound. But then

$$\mathcal{M}^*_{\eta+1} = \mathfrak{C}_{k+1}(\mathcal{M}_b) = \mathfrak{C}_{k+1}(\mathcal{M}_c) = \mathcal{M}^*_{\gamma+1}\,.$$

This implies that $\mathcal{T}$-pred$(\eta+1) = \mathcal{T}$-pred$(\gamma+1)$. For let $\beta = \mathcal{T}$-pred$(\eta+1)$; then E_β is on the $\mathcal{M}^*_{\eta+1}$ sequence, so E_β is on the $\mathcal{M}^*_{\gamma+1}$ sequence, so E_β is on the $\mathcal{M}_\xi$-sequence where $\xi = \mathcal{T}$-pred$(\gamma+1)$. Thus $\xi \leq \beta$ by remark (a) following 5.1. That $\beta \leq \xi$ is proved symmetrically.

Now then

$$i_{\eta+1,b} \circ i^*_{\eta+1} = i_{\gamma+1,c} \circ i^*_{\gamma+1}\,,$$

since by lemma 4.7 each side is the natural embedding from $\mathfrak{C}_{k+1}(\mathcal{M}_b)$ to $\mathfrak{C}_k(\mathcal{M}_b) = \mathcal{M}_b$ inverting the collapse.

Since $\mathcal{T}$ is non-overlapping, crit $i_{\eta+1,b} \geq \rho_\eta$ and crit $i_{\gamma+1,b} \geq \rho_\gamma$. So letting $\nu = \inf(\rho_\eta, \rho_\gamma)$, we have crit $E_\eta =$ crit $E_\gamma < \nu$ and $E_\eta \restriction \nu = E_\gamma \restriction \nu$. By remark (a) following 5.1 we see that $\eta = \gamma$.

Now let β be largest in $b \cap c$; from the above we know that there's no dropping after β on b or c, that is, $\eta+1 = \gamma+1 \in b \cap c$. Let

$$\rho = \sup\{\mathrm{lh}\, E_\xi \mid \xi+1 \in b \cap c\}\,;$$

then

$$\mathcal{M}_\beta = \mathcal{H}^{\mathcal{M}_\beta}_{k+1}(\rho \cup \{q_\beta\})$$

where for any $\xi \in b \cup c$ such that $\xi \geq \eta+1$

$$q_\xi = i_{\eta+1,\xi} \circ i^*_{\eta+1}(p_{k+1}(\mathcal{M}^*_{\eta+1}))\,.$$

But then

$$i_{\beta,b} = i_{\beta,c}\,,$$

as $i_{\beta,b} \restriction \rho = i_{\beta,c} \restriction \rho = \text{id}$, and $i_{\beta,b}(q_\beta) = i_{\beta,c}(q_\beta) = \langle r, u\rangle$, where r is the $k+1$st standard parameter of $(\mathcal{M}_b, u)$ and u is as in the definition of $p_{k+1}(\mathcal{M}_b)$ (cf. Lemma 4.7). Let $\sigma+1 \in b$, $\tau+1 \in c$, and T-pred$(\sigma+1) = T$-pred$(\tau+1) = \beta$. As $i_{\beta,c} = i_{\beta,b}$, we see that crit $E_\sigma =$ crit E_τ, and $E_\sigma \restriction \nu = E_\tau \restriction \nu$, where $\nu = \inf(\rho_\sigma, \rho_\tau)$. This implies $\sigma = \tau$, a contradiction. □

In view of Claims 3 and 4, we may assume there's no dropping of any kind along b (perhaps by exchanging b for c). The proof of the following claim will take several pages and will nearly finish the proof of theorem 6.2.

Claim 5. $\rho_{n+1}^{\mathcal{M}_b} < \delta$.

Proof. We show by induction on $\eta \in b$, that if $\alpha T \eta$, or if $\eta = b$ and $\alpha \in b$, then

$$(*) \qquad \rho_{n+1}^{\mathcal{M}_\eta} \leq i_{\alpha,\eta}(\rho_{n+1}^{\mathcal{M}_\alpha}),$$

and

$$(**) \quad \text{If } \rho_{n+1}^{\mathcal{M}_\eta} = i_{\alpha\eta}(\rho_{n+1}^{\mathcal{M}_\alpha}) \text{ and } \mathrm{Th}_{n+1}^{\mathcal{M}_\alpha}(\rho_{n+1}^{\mathcal{M}_\alpha} \cup \{q\} \notin \mathcal{M}_\alpha)$$
$$\text{then } \mathrm{Th}_{n+1}^{\mathcal{M}_\eta}(\rho_{n+1}^{\mathcal{M}_\eta} \cup \{i_{\alpha\eta}(q)\}) \notin \mathcal{M}_\eta.$$

By (*) for $\eta = b$ and $\alpha = 0$ we have $\rho_{n+1}^{\mathcal{M}_b} \leq i_{0b}(\rho_{n+1}^{\mathcal{M}_0}) \leq \mathrm{lh}\, E$ for some extender E from the $\mathcal{M}_b$ sequence, so that $\rho_{n+1}^{\mathcal{M}_b} < \delta$, as desired.

Consider first the case η is a limit or $\eta = b$. Let $\alpha T \eta$ be the least ordinal such that $i_{\alpha\gamma}(\rho_{n+1}^{\mathcal{M}_\alpha}) = \rho_{n+1}^{\mathcal{M}_\gamma}$ whenever $\alpha T \gamma T \eta$. Such an ordinal α exists by (*). It will be enough to show that whenever $\gamma \in [\alpha, \eta)_T$ and $\mathrm{Th}_{n+1}^{\mathcal{M}_\gamma}(\rho_{n+1}^{\mathcal{M}_\gamma} \cup \{q\})$ is not a member of $\mathcal{M}_\gamma$, then

$$\mathrm{Th}_{n+1}^{\mathcal{M}_\eta}(i_{\gamma\eta}(\rho_{n+1}^{\mathcal{M}_\gamma}) \cup \{i_{\gamma\eta}(q)\}) \notin |\mathcal{M}_\eta|.$$

For this, suppose $\mathrm{Th}_{n+1}^{\mathcal{M}_\eta}(i_{\gamma\eta}(\rho_{n+1}^{\mathcal{M}_\gamma}) \cup \{i_{\gamma\eta}(q)\}) = i_{\xi\eta}(x)$, where we may assume $\gamma T \xi T \eta$. As $i_{\xi\eta}$ is generalized $r\Sigma_{n+1}$ elementary, we see $x = \mathrm{Th}_{n+1}^{\mathcal{M}_\xi}(i_{\gamma\xi}(\rho_{n+1}^{\mathcal{M}_\gamma} \cup \{i_{\gamma\xi}(q)\})$. This contradicts (**) at ξ.

Now let $\eta = \xi + 1$ and set $\beta = T$-pred(η). If (*) or (**) fails at η we must have $q \in |\mathcal{M}_\beta|$ such that

$$\mathrm{Th}_{n+1}^{\mathcal{M}_\beta}(\rho_{n+1}^{\mathcal{M}_\beta} \cup \{q\}) \notin |\mathcal{M}_\beta|$$

but

$$\mathrm{Th}_{n+1}^{\mathcal{M}_\eta}(i_{\beta\eta}(\rho_{n+1}^{\mathcal{M}_\beta}) \cup \{i_{\beta\eta}(q)\}) = [a, f]_{E_\xi}^{\mathcal{M}_\beta} \in |\mathcal{M}_\eta|.$$

Fix such a q. Let $\rho = \rho_{n+1}^{\mathcal{M}_\beta}$, $i = i_{\beta\eta}$, $E = E_\xi$.

We may assume $f(\bar{u}) \subseteq \rho$ for all $\bar{u} \in \mathrm{dom}\, f$. Also $\rho < \rho_n^{\mathcal{M}_\beta}$ by (*) and the fact that $\rho_{n+1}^{\mathcal{M}_0} < \rho_n^{\mathcal{M}_0}$. If we let $A = \{(\bar{u}, \nu) \mid \nu \in f(\bar{u})\}$, then A is (generalized) $r\Sigma_n$, so $A \in |\mathcal{M}_\beta|$. Thus $f \in |\mathcal{M}_\beta|$.

Now

$$(\dagger) \qquad x \in \mathrm{Th}_{n+1}^{\mathcal{M}_\beta}(\rho \cup \{q\}) \Leftrightarrow i(x) \in [a, f]_E^{\mathcal{M}_\beta}$$

since i is generalized $r\Sigma_{n+1}$ elementary. This gives an $r\Delta_1^{\mathcal{M}_\beta}$ definition of $\mathrm{Th}_{n+1}^{\mathcal{M}_\beta}(\rho \cup \{q\})$ since E_a is $r\Sigma_1^{\mathcal{M}_\beta}$. This is a contradiction if $n > 0$, so we now assume $n = 0$.

Let $\kappa = \mathrm{crit}\ E$. We have $\kappa < \rho$ by Lemma 4.5. On the other hand, $E_a \notin |\mathcal{M}_\beta|$, as otherwise $(\dagger)$ would imply $\mathrm{Th}_1^{\mathcal{M}_\beta}(\rho \cup \{q\}) \in |\mathcal{M}_\beta|$. Thus $\rho = \rho_1^{\mathcal{M}_\beta} = (\kappa^+)^{\mathcal{M}_\beta}$.

We will now complete the proof of claim 5 by showing that there is a $r\Sigma_1^{\mathcal{M}_\beta}$ function $t : \kappa \to \rho$ such that $\mathrm{ran}(t)$ is cofinal in ρ. To see that this proves claim 5, we let S be the set of triples (α, γ, ν) such that $\gamma \prec_{t(\alpha)} \nu$, where $\prec_{t(\alpha)}$ is the first well ordering of κ in the natural order of $\mathcal{M}_\beta$ which has order type $t(\alpha)$. Then $S \subseteq \kappa$ and S is $r\Sigma_1^{\mathcal{M}_\beta}$, so that $S \in |\mathcal{M}_\beta|$ and hence $\rho < (\kappa^+)^{\mathcal{M}_\beta}$, contradiction.

For any $\mathcal{N}$ and $X \subseteq |\mathcal{N}|$, let

$$\overline{\mathrm{Th}}_1^{\mathcal{N}}(X) = \mathrm{Th}_1^{\mathcal{N}}(X) \cap \{(\varphi, \bar{a}) \mid \varphi \text{ is pure } r\Sigma_1\}.$$

Using the proof of Lemma 2.10 we see that $\mathrm{Th}_1^{\mathcal{M}_\beta}(\rho \cup \{q\}) \notin |\mathcal{M}_\beta|$ implies that $\overline{\mathrm{Th}}_1^{\mathcal{M}_\beta}(\rho \cup \{q\}) \notin |\mathcal{M}_\beta|$, so we can use $\overline{\mathrm{Th}}_1^{\mathcal{M}_\beta}(\rho \cup \{q\})$ instead of $\mathrm{Th}_1^{\mathcal{M}_\beta}(\rho \cup \{q\})$. Let f be the function representing $\overline{\mathrm{Th}}_1^{\mathcal{M}_\eta}(i(\rho) \cup \{i(q)\})$. We need to consider two cases:

Case 1. There is a total, continuous, order-preserving, $r\Sigma_1^{\mathcal{M}_\beta}$ function $g : \kappa \to \mathrm{OR}^{\mathcal{M}_\beta}$ such that $g''\kappa$ is cofinal in $\mathrm{OR}^{\mathcal{M}_\beta}$.

In this case, we set for $\bar{u} \in \mathrm{dom}(f)$

$$h(\bar{u}) = \overline{\mathrm{Th}}_1^{J_{g(u_0)}^{\mathcal{M}_\beta}}(\rho \cup \{q\}),$$

so that h is $r\Sigma_1^{\mathcal{M}_\beta}$. Notice that if $A \in E_a$, then $\exists \bar{u} \in A\, h(\bar{u}) \neq f(\bar{u})$, as otherwise $h \restriction A \in |\mathcal{M}_\beta|$, so that $\overline{\mathrm{Th}}_1^{\mathcal{M}_\beta}(\rho \cup \{q\}) \in |\mathcal{M}_\beta|$, a contradiction.

Now set, for all $\bar{u} \in \mathrm{dom}(f)$

$$t(\bar{u}) = \begin{cases} \text{least } \alpha & \text{such that } (f(\bar{u}) \triangle h(\bar{u})) \cap (\omega \times (\alpha \cup \{q\})^{<\omega}) \neq \emptyset \\ 0 & \text{if no such } \alpha \text{ exists}. \end{cases}$$

So t is total and $r\Sigma_1^{\mathcal{M}_\beta}$. It is enough to see $\mathrm{ran}(t)$ is unbounded in ρ. Fix any ordinal $\theta < \rho$. We will complete the proof of case 1 by finding a $\bar{u}$ such that $t(\bar{u}) > \theta$. Define a function k by

$$k(\bar{v}) = h(\bar{v}) \cap (\omega \times (\theta \cup \{q\})^{<\omega}).$$

Then $k \in |\mathcal{M}_\beta|$ since it can be computed from $\mathrm{Th}_1^{\mathcal{M}_\beta}(\theta \cup \{q, r\})$, where r is a parameter chosen so that the function g is $\Sigma_1^{\mathcal{M}_\beta}(\{r\})$. Moreover

$$(\dagger\dagger) \qquad [a, k]_E^{\mathcal{M}_\beta} = \overline{\mathrm{Th}}_1^{\mathcal{M}_\eta}(i(\theta) \cup \{i(q)\}).$$

One direction, $\supseteq$, of equation $(\dagger\dagger)$ is easy. To prove $\subseteq$, let $[b, \mathcal{I}]_E^{\mathcal{M}_\beta} \in [a, k]_E^{\mathcal{M}_\beta}$, where we may assume $a \subseteq b$. We may assume that for all $\bar{v} \in \operatorname{dom} \mathcal{I}$

$$\mathcal{I}(\bar{v}) \in k(\bar{v}^*) = \overline{\mathrm{Th}}_1^{J^{\mathcal{M}_\beta}_{g(v_0)}}(\theta \cup \{q\})$$

where $\bar{v}^*$ is the appropriate subsequence of $\bar{v}$. For $\bar{v} \in \operatorname{dom} \mathcal{I}$ such that v_0 is a limit, let

$$s(\bar{v}) = \text{least } \alpha < v_0 \text{ such that } \mathcal{I}(\bar{v}) \in \overline{\mathrm{Th}}_1^{J^{\mathcal{M}_\beta}_{g(\alpha)}}(\theta \cup \{q\}).$$

Then s is a $r\Sigma_1^{\mathcal{M}_\beta}$ map from κ^n to κ, so $s \in |\mathcal{M}_\beta|$. By normality, fix α_0 such that $s(\bar{v}) = \alpha_0$ for E_b a.e. $\bar{v}$, and let $\xi = g(\alpha_0)$. Then

$$\begin{aligned} [a, \mathcal{I}]_E^{\mathcal{M}_\beta} &\in i(\overline{\mathrm{Th}}_1^{J^{\mathcal{M}_\beta}_\xi}(\theta \cup \{q\})) \\ &= \overline{\mathrm{Th}}_1^{J^{\mathcal{M}_\eta}_{i(\xi)}}(i(\theta) \cup \{i(q)\}) \subseteq \overline{\mathrm{Th}}_1^{\mathcal{M}_\eta}(i(\theta) \cup \{i(q)\}), \end{aligned}$$

as desired. This completes the proof of equation $(\dagger\dagger)$.

It follows that there is an $A \in E_a$ such that for all $\bar{u} \in A$,

$$f(\bar{u}) \cap (\omega \times (\theta \cup \{q\})^{<\omega}) = h(\bar{u}) \cap (\omega \times (\theta \cup \{q\})^{<\omega}).$$

Let $\bar{u} \in A$ be such that $h(\bar{u}) \neq f(\bar{u})$; then $t(\bar{u}) > \theta$. This completes the proof of case 1 of claim 5.

Case 2. There is no function g as in case 1.

In this case, define the function $t(\bar{u})$, where $\bar{u} \in \operatorname{dom}(f)$, by

$$t(\bar{u}) = \text{least } \alpha \text{ such that } \left(f(\bar{u}) \,\triangle\, \overline{\mathrm{Th}}_1^{\mathcal{M}_\beta}(\rho \cup \{q\})\right) \cap (\omega \times (\alpha \cup \{q\})^{<\omega}) \neq \varnothing.$$

Thus t is total $r\Sigma_1^{\mathcal{M}_\beta}$. To see that $\operatorname{ran} t$ is unbounded in ρ, note that for $\theta < \rho$

$$\overline{\mathrm{Th}}_1^{\mathcal{M}_\eta}(i(\theta) \cup \{i(q)\}) = i(\overline{\mathrm{Th}}_1^{\mathcal{M}_\beta}(\theta \cup \{q\}))$$

as

$$\overline{\mathrm{Th}}_1^{\mathcal{M}_\beta}(\theta \cup \{q\}) = \overline{\mathrm{Th}}_1^{J^{\mathcal{M}_\beta}_\xi}(\theta \cup \{q\})$$

for some $\xi < \mathrm{OR}^{\mathcal{M}_\beta}$ by case hypothesis.

This completes the proof of case 2, and hence of Claim 5. □

Fix now $p \in |\mathcal{M}_\beta|$ and $\rho < \delta$ such that $\mathrm{Th}_{n+1}^{\mathcal{M}_b}(\rho \cup \{p\}) \notin |\mathcal{M}_b|$. We obtain a contradiction via an easy generalization of the proof of 6.1.

Fix $\beta <$ length of $\mathcal{T}$ so large that

(1) $b \cap \beta \neq c \cap \beta$, and there's no dropping on $b \cup c$ above β.

(2) $\gamma \in b - \beta \Rightarrow \mathrm{crit}\ i_{\gamma b} > \rho$ and $p \in \mathrm{ran}\ i_{\gamma b}$ and ($\delta < \mathrm{OR}^{\mathcal{M}_b} \Rightarrow \delta \in \mathrm{ran}\ i_{\gamma b}$).

(3) $\gamma \in c - \beta \Rightarrow \mathrm{crit}\ i_{\gamma c} > \rho$ and $p \in \mathrm{ran}\ i_{\gamma c}$ and ($\delta < \mathrm{OR}^{\mathcal{M}_b} \Rightarrow \delta \in \mathrm{ran}\ i_{\gamma c}$) and ($\mathrm{OR}^{\mathcal{M}_b} < \mathrm{OR}^{\mathcal{M}_c} \Rightarrow \mathrm{OR}^{\mathcal{M}_b} \in \mathrm{ran}\ i_{\gamma c}$).

As in Claim 2 of the proof of 6.1, we can find $\gamma \in b - \beta$ and $\eta \in c - \beta$ such that

$$\mathrm{ran}\ i_{\gamma b} \cap \mathrm{ran}\ i_{\eta c} \cap \delta = \kappa$$

where $\rho < \kappa < \delta$. Let

$$\pi : |\mathcal{N}| \cong X \subseteq |\mathcal{M}_b|$$

where $X = \mathrm{ran}\ i_{\gamma b} \cap \mathrm{ran}\ i_{\eta c}$ and π is the inverse of the collapse. Then π is generalized $r\Sigma_{n+1}$ elementary. This follows from the fact that both $i_{\gamma b}$ and $i_{\gamma c}$ are generalized $r\Sigma_{n+1}$ elementary. To see that $i_{\gamma c}$ is generalized $r\Sigma_{n+1}$ elementary, note that if $\mathcal{M}_b = \mathcal{M}_c$, then $\deg(\xi+1) \geq n$ for all sufficiently large $\xi + 1 \in c$, so $i_{\eta c}$ is generalized $r\Sigma_{n+1}$ elementary. If $\mathcal{M}_b$ is a proper initial segment of $\mathcal{M}_c$, then $i_{\eta c} \upharpoonright i_{\eta c}^{-1}(\mathcal{M}_b)$ is in fact fully elementary.

Notice that $\mathrm{crit}\ \pi = \kappa$, and $\mathcal{N} = \mathcal{J}_\alpha^{\vec{E}(\mathcal{T})\upharpoonright\kappa}$ for some $\alpha \geq \kappa$. Also $\mathrm{Th}_{n+1}^{\mathcal{M}_b}(\rho \cup \{p\})$ is definable over $\mathcal{N}$, and hence is a member of $L[\vec{E}(\mathcal{T}) \upharpoonright \kappa]$. As $\vec{E}(\mathcal{T}) \upharpoonright \kappa \in |\mathcal{M}_b|$ and $\mathcal{M}_b$ has an internally iterable extender on its sequence with critical point greater than κ, we get $\mathrm{Th}_{n+1}^{\mathcal{M}_b}(\rho \cup \{p\}) \in |\mathcal{M}_b|$, a contradiction. This completes the proof of theorem 6.2. □

§7. The Comparison Process

We prove in this section a comparison lemma for 1-small mice. Our interest is not so much in the lemma itself, but in the method by which it is proved. We shall use that method in a much more important way in the next section.

For bookkeeping purposes we shall use "padded iteration trees". These are just like ordinary iteration trees except that we modify the successor clause in the definition of "iteration tree" so as to allow $\alpha T(\alpha+1)$, $\mathcal{M}_\alpha = \mathcal{M}_{\alpha+1}$, and $i_{\alpha,\alpha+1} =$ identity, and then require that $\alpha T \beta \Rightarrow \beta = \alpha+1$ or $(\alpha+1)\, T \beta$. So a padded tree is essentially an ordinary tree with the indexing of the models slowed down by repetition. We shall no doubt often fail to distinguish between iteration trees and their padded counterparts.

Theorem 7.1 (The comparison lemma). *Let $\mathcal{M}$ and $\mathcal{N}$ be n-sound, 1-small, n-iterable premice, where $n \leq \omega$. Then there are n-maximal padded iteration trees $\mathcal{T}$ on $\mathcal{M}$ and $\mathcal{U}$ on $\mathcal{N}$ such that either*

(1) *$\mathcal{T}$ and $\mathcal{U}$ have successor length $\theta+1$, and either*

(a) *$\mathcal{M}_\theta$ is an initial segment of $\mathcal{N}_\theta$ and $D^{\mathcal{T}} \cap [0,\theta]_T = \varnothing$ and deg$(\alpha+1) = n$ for all $\alpha+1 \in [0,\theta]_T$, or*

(b) *$\mathcal{N}_\theta$ is an initial segment of $\mathcal{M}_\theta$ and $D^{\mathcal{U}} \cap [0,\theta]_U = \varnothing$ and deg$(\alpha+1) = n$ for all $\alpha+1 \in [0,\theta]_U$,*

or

(2) *$\mathcal{T}$ and $\mathcal{U}$ have limit length, one of the two is not simple, and in some $V^{\mathrm{Col}(\kappa,\omega)}$ there are wellfounded cofinal branches b of $\mathcal{T}$ and c of $\mathcal{U}$ such that either*

(a) *$\mathcal{M}_b$ is an initial segment of $\mathcal{N}_c$, $D^{\mathcal{T}} \cap b = \varnothing$, and deg$(\alpha+1) = n$ for all $\alpha+1 \in b$, or*

(b) *$\mathcal{N}_c$ is an initial segment of $\mathcal{M}_b$, $D^{\mathcal{U}} \cap c = \varnothing$, and deg$(\alpha+1) = n$ for all $\alpha+1 \in c$.*

Proof. We define by induction on γ

$$\mathcal{T} \upharpoonright \gamma = \langle T \cap (\gamma \times \gamma), D^{\mathcal{T}} \cap \gamma, \deg^{\mathcal{T}} \upharpoonright \gamma, \langle E^{\mathcal{T}}_\alpha, \mathcal{M}^*_{\alpha+1} \mid \alpha+1 < \gamma \rangle\rangle$$

and

$$\mathcal{U} \upharpoonright \gamma = \langle U \cap (\gamma \times \gamma), D^{\mathcal{U}} \cap \gamma, \deg^{\mathcal{U}} \upharpoonright \gamma, \langle E^{\mathcal{U}}_\alpha, \mathcal{N}^*_{\alpha+1} \mid \alpha+1 < \gamma \rangle\rangle$$

together with the associated $\mathcal{M}_\alpha$, $\mathcal{N}_\alpha$ for $\alpha < \gamma$, $\rho^{\mathcal{T}}_\alpha$ and $\rho^{\mathcal{U}}_\alpha$ for $\alpha+1 < \gamma$, and embeddings $i^{\mathcal{T}}_{\alpha\beta}$, $i^{\mathcal{U}}_{\alpha\beta}$ (for (α,β) as appropriate). The method for defining $\mathcal{T}$ and $\mathcal{U}$ is the standard one of "iterating the least disagreement".

We begin with $\gamma = 1$. In this case we need only define $\mathcal{M}_0$ and $\mathcal{N}_0$, which we do by setting

$$\mathcal{M}_0 = \mathcal{M}, \qquad \mathcal{N}_0 = \mathcal{N}.$$

Now consider the case γ is a limit ordinal. Then

$$\mathcal{T} \restriction \gamma = \text{``}\bigcup_{\beta<\gamma}\text{''}\, \mathcal{T} \restriction \beta$$
$$\mathcal{U} \restriction \gamma = \text{``}\bigcup_{\beta<\gamma}\text{''}\, \mathcal{U} \restriction \beta$$

where the union is taken along each of the 4 coordinates.

Now suppose $\gamma = \lambda + 1$, where λ is a limit ordinal. If $\mathcal{T} \restriction \lambda$ or $\mathcal{U} \restriction \lambda$ is not simple, we stop our induction. Suppose $\mathcal{T} \restriction \lambda$ and $\mathcal{U} \restriction \lambda$ are simple. As $\mathcal{M}$ and $\mathcal{N}$ are n-iterable we have wellfounded branches b of $\mathcal{T} \restriction \lambda$ and c of $\mathcal{U} \restriction \lambda$ which are cofinal in λ. Set

$$\mathcal{M}_\lambda = \mathcal{M}_b, \qquad \mathcal{N}_\lambda = \mathcal{N}_c,$$

$$T \cap (\gamma \times \gamma) = (T \cap (\lambda \times \lambda)) \cup \{(\alpha, \lambda) \mid \alpha \in b\}$$
$$U \cap (\gamma \times \gamma) = (U \cap (\lambda \times \lambda)) \cup \{(\alpha, \lambda) \mid \alpha \in c\},$$

$$i^{\mathcal{T}}_{\alpha\lambda} = i^{\mathcal{T}}_{\alpha b} \quad \text{for } \alpha \in b - \sup(D^{\mathcal{T}} \cap b),$$
$$i^{\mathcal{U}}_{\alpha\lambda} = i^{\mathcal{U}}_{\alpha c} \quad \text{for } \alpha \in c - \sup(D^{\mathcal{N}} \cap c).$$

The rest of $\mathcal{T} \restriction \gamma$ and $\mathcal{U} \restriction \gamma$ is determined by this.

Finally, we have the case $\gamma = \eta+2$. Here we must define $E^{\mathcal{T}}_\eta$, $\mathcal{M}^*_{\eta+1}$, $D^{\mathcal{T}} \cap \{\eta+1\}$, $\deg^{\mathcal{T}}(\eta+1)$, T-pred$(\eta+1)$, and similarly for the $\mathcal{U}$ side. We are given the models $\mathcal{M}_\eta$ and $\mathcal{N}_\eta$.

If $\mathcal{M}_\eta$ is an initial segment of $\mathcal{N}_\eta$, or vice-versa, then we stop our inductive definition. Otherwise we have a least ordinal γ such that $\mathcal{J}^{\mathcal{M}_\eta}_\gamma \neq \mathcal{J}^{\mathcal{N}_\eta}_\gamma$. Set

$$E^{\mathcal{T}}_\eta = \begin{cases} \varnothing & \text{if } \mathcal{J}^{\mathcal{M}_\eta}_\gamma \text{ is passive}, \\ \dot{F}^{\mathcal{J}^{\mathcal{N}_\eta}_\gamma} & \text{if } \mathcal{J}^{\mathcal{N}_\eta}_\gamma \text{ is active}, \end{cases}$$

$$E^{\mathcal{U}}_\eta = \begin{cases} \varnothing & \text{if } \mathcal{J}^{\mathcal{N}_\eta}_\gamma \text{ is passive}, \\ \dot{F}^{\mathcal{J}^{\mathcal{N}_\eta}_\gamma} & \text{if } \mathcal{J}^{\mathcal{N}_\eta}_\gamma \text{ is active}. \end{cases}$$

The rest is determined by the rules for forming non-overlapping, n-maximal, padded iteration trees. So, on the $\mathcal{T}$ side:

If $E_\eta^{\mathcal{T}} = \emptyset$, then T-pred$(\eta+1) = \eta$, $\mathcal{M}_{\eta+1}^* = \mathcal{M}_{\eta+1} = \mathcal{M}_\eta$, $i_{\eta,\eta+1}^{\mathcal{T}} =$ identity. In this case, we also set $\rho_\eta^{\mathcal{T}} = 0$.

Now suppose $E_\eta^{\mathcal{T}} \neq \emptyset$. Let $\kappa =$ crit $E_\eta^{\mathcal{T}}$, and let $\beta \leq \eta$ be least such that $\kappa < \rho_\beta^{\mathcal{T}}$. Then we set T-pred$(\eta+1) = \beta$. Let

$$\begin{aligned} \mathcal{M}_{\eta+1}^* &= \text{longest initial segment } \mathcal{P} \text{ of } \mathcal{M}_\beta \text{ such that} \\ &\qquad P(\kappa) \cap |\mathcal{P}| = P(\kappa) \cap |\mathcal{J}_\gamma^{\mathcal{M}_\eta}| \\ &= \text{longest initial segment } \mathcal{P} \text{ of } \mathcal{M}_\beta \text{ such that} \\ &\qquad P(\kappa) \cap |\mathcal{P}| = P(\kappa) \cap |\mathcal{J}_{\mathrm{lh}\, E_\beta}^{\mathcal{M}_\beta}|. \end{aligned}$$

We let

$$\eta + 1 \in D^{\mathcal{T}} \Leftrightarrow \mathcal{M}_{\eta+1}^* \neq \mathcal{M}_\beta .$$

Let

$$\begin{aligned} k &= \text{largest } m \leq \omega \text{ such that } \kappa < \rho_m^{\mathcal{M}_{\eta+1}^*} \text{ and} \\ &\qquad D^{\mathcal{T}} \cap [0, \eta+1]_T = \emptyset \Rightarrow m \leq n . \end{aligned}$$

We let $\deg^{\mathcal{T}}(\eta+1) = k$, and

$$\mathcal{M}_{\eta+1} = \mathrm{Ult}_k(\mathcal{M}_{\eta+1}^*, E_\eta^{\mathcal{T}}),$$

and let $i_{\eta+1}^* : \mathcal{M}_{\eta+1}^* \to \mathcal{M}_{\eta+1}$ be the canonical embedding, and for $\alpha T(\eta+1)$ such that $D^{\mathcal{T}} \cap (\alpha, \eta+1]_T = \emptyset$, let $i_{\alpha,\eta+1}^{\mathcal{T}} = i_{\eta+1}^* \circ i_{\alpha,\beta}^{\mathcal{T}}$.

This completes the definition of $\mathcal{T} \restriction \eta+2$. We obtain $\mathcal{U} \restriction \eta+2$ from $E_\eta^{\mathcal{U}}$ in a similar fashion.

This completes the definitions of $\mathcal{T}$ and $\mathcal{U}$. It is easy to see they are iteration trees.

Claim. If $\alpha < \beta$, then $\max(\rho_\alpha^{\mathcal{T}}, \rho_\alpha^{\mathcal{U}}) < \min(\rho_\beta^{\mathcal{T}}, \rho_\beta^{\mathcal{U}})$.

Proof. $\gamma = \mathrm{lh}\, E_\alpha^{\mathcal{T}}$ is a cardinal of $\mathcal{M}_\beta$, and hence a cardinal of $J_{\mathrm{lh}\, E_\beta^{\mathcal{N}}}^{\mathcal{N}_\beta}$. As γ is a cardinal of $J_{\mathrm{lh}\, E_\beta^{\mathcal{T}}}^{\mathcal{M}_\beta}$, $\gamma \leq \rho_\beta^{\mathcal{T}}$. As γ is a cardinal of $J_{\mathrm{lh}\, E_\beta^{\mathcal{T}}}^{\mathcal{N}_\beta}$, $\gamma \leq \rho_\beta^{\mathcal{U}}$. So $\rho_\alpha^{\mathcal{T}} < \gamma \leq \min(\rho_\beta^{\mathcal{T}}, \rho_\beta^{\mathcal{U}})$. Symmetrically, $\rho_\alpha^{\mathcal{U}} < \min(\rho_\beta^{\mathcal{T}}, \rho_\beta^{\mathcal{U}})$.

Lemma 7.2. *Let $\alpha+1, \beta+1 < \mathrm{lh}\,\mathcal{T}$. Suppose $E_\alpha^{\mathcal{T}} \neq \emptyset$ and $E_\beta^{\mathcal{U}} \neq \emptyset$. Suppose crit $E_\alpha^{\mathcal{T}} =$ crit $E_\beta^{\mathcal{U}} = \kappa$. Then there is a parameter $a \in [\rho_\alpha^{\mathcal{T}} \cap \rho_\beta^{\mathcal{U}}]^{<\omega}$, and a set $A \subseteq [\kappa]^{\mathrm{card}\, a}$ such that*

$$A \in (E_\alpha^{\mathcal{T}})_a \quad \textit{and} \quad A \notin (E_\beta^{\mathcal{U}})_a .$$

PROOF. We may as well assume $\alpha \leq \beta$. Notice then $\rho_\alpha^{\mathcal{T}} \leq \rho_\beta^{\mathcal{U}}$, and

$$P(\kappa) \cap J^{\mathcal{M}_\alpha}_{\mathrm{lh}\, E_\alpha^{\mathcal{T}}} = P(\kappa) \cap J^{\mathcal{M}_\beta}_{\mathrm{lh}\, E_\beta^{\mathcal{U}}} = P(\kappa) \cap J^{\mathcal{N}_\beta}_{\mathrm{lh}\, E_\beta^{\mathcal{U}}}$$

by 5.1 and the fact that $\mathcal{M}_\beta$ and $\mathcal{N}_\beta$ agree below $\mathrm{lh}\, E_\beta^{\mathcal{U}}$. So $E_\alpha^{\mathcal{T}}$ and $E_\beta^{\mathcal{U}}$ are defined on the same subsets of κ, and it will suffice to show

$$E_\alpha^{\mathcal{T}} \restriction \rho_\alpha^{\mathcal{T}} \neq E_\beta^{\mathcal{U}} \restriction \rho_\alpha^{\mathcal{T}}.$$

Suppose $E_\alpha^{\mathcal{T}} \restriction \rho_\alpha^{\mathcal{T}} = E_\beta^{\mathcal{U}} \restriction \rho_\alpha^{\mathcal{T}}$. Now $E_\alpha^{\mathcal{T}}$ is the trivial completion of $E_\alpha^{\mathcal{T}} \restriction \rho_\alpha^{\mathcal{T}}$, and so the initial segment condition on $\mathcal{N}_\beta$ gives us two possibilities. We may have $E_\alpha^{\mathcal{T}}$ on the sequence of $\mathcal{N}_\beta$ at or before the position of $E_\beta^{\mathcal{U}}$. But $E_\alpha^{\mathcal{T}}$ is not on the $\mathcal{N}_\alpha$ sequence because it is part of the least disagreement at α, and by coherence then $E_\alpha^{\mathcal{T}}$ is not on the $\mathcal{N}_\beta$ sequence for all $\beta \geq \alpha$. Thus the second possibility from the initial segment condition is realized: $\rho_\alpha^{\mathcal{T}} \in \mathrm{dom}\, \dot{E}^{\mathcal{N}_\beta}$ and $E_\alpha^{\mathcal{T}}$ is on the sequence of $\mathrm{Ult}_0(\mathcal{P}, F)$, where $F = (\dot{E}^{\mathcal{N}_\beta})_{\rho_\alpha^{\mathcal{T}}}$ and $\mathcal{P} = J^{\mathcal{N}_\beta}_{\rho_\alpha^{\mathcal{T}}}$. In this case, F is on the sequence of $\mathcal{N}_\beta$, and hence of $\mathcal{M}_\beta$ as $\rho_\alpha^{\mathcal{T}} < \mathrm{lh}\, E_\beta^{\mathcal{U}}$. Thus F is on the sequence of $\mathcal{M}_\alpha$ as $\rho_\alpha^{\mathcal{T}} < \mathrm{lh}\, E_\alpha^{\mathcal{T}}$. Also $\mathcal{P} = J^{\mathcal{M}_\alpha}_{\rho_\alpha^{\mathcal{T}}}$. By coherence, F is on the sequence of $\mathrm{Ult}_0(\mathcal{P}, E_\alpha^{\mathcal{T}})$ and F is not on the sequence of $\mathrm{Ult}_0\, (\mathrm{Ult}_0(\mathcal{P}, F), E_\alpha^{\mathcal{T}})$. This is a contradiction, as these ultrapowers agree past $\mathrm{lh}\, E_\alpha^{\mathcal{T}}$. □

CLAIM. The inductive definitions of $\mathcal{T}$ and $\mathcal{U}$ halt at some ordinal γ such that $\gamma \leq \max(\mathrm{card}\, \mathcal{M},\ \mathrm{card}\, \mathcal{N})^+$.

PROOF. Let $\theta = \max(\mathrm{card}\, \mathcal{M},\ \mathrm{card}\, \mathcal{N})^+$. If the claim is false, then $\mathcal{M}_\theta$ and $\mathcal{N}_\theta$ are defined. Let $b = [0,\theta)_T$ and $c = [0,\theta)_U$. So b and c are club in θ. By the standard closure argument we can find a club $d \subseteq b \cup c$ such that

(i) $D^{\mathcal{T}} \cap d = D^{\mathcal{N}} \cap d = \emptyset$.

(ii) $\alpha \in d \Rightarrow \alpha = \mathrm{crit}\ i^{\mathcal{T}}_{\alpha b} = \mathrm{crit}\ i^{\mathcal{U}}_{\alpha c}$.

(iii) $\alpha, \beta \in d \wedge \alpha < \beta \Rightarrow (i^{\mathcal{T}}_{\alpha\beta}(\alpha) = \beta \wedge i^{\mathcal{U}}_{\alpha\beta}(\alpha) = \beta)$.

(iv) $(\alpha \in d \wedge A \subseteq [\alpha]^n \wedge A \in |\mathcal{M}_\alpha| \cap |\mathcal{N}_\alpha|) \Rightarrow i^{\mathcal{T}}_{\alpha b}(A) = i^{\mathcal{U}}_{\alpha c}(A)$.

Now let d satisfy (i)-(iv) and take $\alpha \in d$. Let $\beta + 1$ and $\gamma + 1$ be the successor of α in b and c respectively, so that T-$\mathrm{Pred}(\beta+1) = U$-$\mathrm{Pred}(\gamma+1) = \alpha$. Since $\mathcal{T}$ and $\mathcal{U}$ are non-overlapping, $\mathrm{crit}\ i^{\mathcal{T}}_{\beta+1,b} \geq \rho^{\mathcal{T}}_\beta$ and $\mathrm{crit}\ i^{\mathcal{U}}_{\gamma+1,c} \geq \rho^{\mathcal{U}}_\gamma$. By (iv) we see that for all $A \subseteq [\alpha]^n$, $A \in |\mathcal{M}_\alpha| \cap |\mathcal{N}_\alpha|$,

$$i^{\mathcal{T}}_{\alpha,\beta+1}(A) \cap [\rho]^n = i^{\mathcal{U}}_{\alpha,\gamma+1}(A) \cap [\rho]^n$$

where $\rho = \rho^{\mathcal{T}}_\beta \cap \rho^{\mathcal{U}}_\gamma$. It follows that $E^{\mathcal{T}}_\beta \restriction \rho = E^{\mathcal{U}}_\gamma \restriction \rho$, contradicting the lemma. This proves our claim. □

There are two ways the construction of $\mathcal{T}$ and $\mathcal{U}$ can halt. Suppose first we reach $\theta + 1$ such that $\mathcal{M}_\theta$ is an initial segment of $\mathcal{N}_\theta$ or vice-versa. If $\mathcal{M}_\theta$ is a proper initial segment of $\mathcal{N}_\theta$, then there's no dropping along $[0,\theta]_T$ because $\mathcal{N}_\theta$ is a premouse so that $\mathcal{M}_\theta$ is ω-sound. So we have (1) (a) of our desired conclusion. If $\mathcal{N}_\theta$ is a proper initial segment of $\mathcal{M}_\theta$ we have (1) (b) of our desired conclusion. Finally, if $\mathcal{M}_\theta = \mathcal{N}_\theta$ then on one of $[0,\theta]_T$ and $[0,\theta]_U$ there's no dropping; the proof is just like that of Claim 4 in the proof of 6.2, so we omit it.

Suppose next the construction halts because we reach a limit ordinal θ such that one of $\mathcal{T} \restriction \theta$ and $\mathcal{U} \restriction \theta$ is not simple. Say $\mathcal{T} \restriction \theta$ is not simple, so there are distinct wellfounded cofinal branches of $\mathcal{T} = \mathcal{T} \restriction \theta$. Just as in the proof of the first 4 claims of 6.2, we can find a cofinal wellfounded branch b of $\mathcal{T}$ such that $D^\mathcal{T} \cap b = \emptyset$ and $\deg(\alpha+1) = n$ for all $\alpha + 1 \in b$, and $\mathcal{M}_b$ has no extenders with length $\geq \delta = \delta(\mathcal{T})$. Let c be any cofinal, wellfounded branch of $\mathcal{U}$. If $\mathcal{M}_b$ is an initial segment of $\mathcal{N}_c$, we are done. If $\mathcal{N}_c$ is a proper initial segment of $\mathcal{M}_b$ then $\mathcal{N}_c$ is ω-sound, so there's no dropping along c and we're done. The remaining possibility (since $\mathcal{M}_b$ and $\mathcal{N}_c$ agree below δ) is that $\mathcal{N}_c$ has an extender F such that $\delta \leq \operatorname{lh} F \leq \mathrm{OR}^{\mathcal{M}_b}$. But this contradicts the 1-smallness of $\mathcal{N}_c$. □

Remark. We haven't ruled out the possibility that (1) of our Theorem 7.1 holds, that $\mathcal{M}_\theta = \mathcal{N}_\theta$, and that one (but not both!) of $[0,\theta]_T$ and $[0,\theta]_U$ has a drop. Nor have we ruled out the analogous situation in case (2) of 7.1. One can show that this cannot happen in the case $n = \omega$, but for $n < \omega$ we don't know.

§8. SOLIDITY AND CONDENSATION

In this section we prove the central fine structural result of the theory we are developing, namely that every 1-small mouse is k-solid for all k. We also derive, by the same method, some condensation results we shall need later. Our proofs of these facts trace back to Dodd's proof that the models of [D] satisfy the GCH.

For mice $\mathcal{M}$ up to a strong cardinal (that is, for mice $\mathcal{M}$ such that $\mathcal{J}_\kappa^{\mathcal{M}} \models$ "There are no strong cardinals" whenever $\kappa = \operatorname{crit} E$ for some extender E on the $\mathcal{M}$ sequence), our proof actually shows that $\mathfrak{C}_k(\mathcal{M})$ is an iterate of $\mathfrak{C}_{k+1}(\mathcal{M})$, with the iteration map having critical point $\geq \rho_{k+1}(\mathcal{M})$. That is, every "very small" mouse is an iterate of its core. We suspect that this is not true for arbitrary 1-small mice.

Recall that $u_0(\mathcal{M}) = \emptyset$, and that $u_k(\mathcal{M}) = \langle \rho_k(\mathcal{M}), b_0, \cdots, b_S, \rho_{k-1}^{\mathcal{M}} \rangle$ for $k \geq 1$, where $b_0 \cdots b_S$ are the solidity witnesses for $p_k(\mathcal{M})$ and the last coordinate $\rho_{k-1}^{\mathcal{M}}$ occurs only if it is defined and is smaller than $\mathrm{OR}^{\mathcal{M}}$. Thus $p_{k+1}(\mathcal{M})$ is the appropriate collapse of $\langle r, u_k(\mathcal{M}) \rangle$, where r is the $k+1$st standard parameter of $(\mathfrak{C}_k(\mathcal{M}), u_k(\mathcal{M}))$.

Recall that if $\pi : \mathcal{M} \to \mathcal{N}$ is a k-embedding, then $\pi(u_k(\mathcal{M})) = u_k(\mathcal{N})$.

Theorem 8.1. *Let $\mathcal{M}$ be a k-sound, 1-small, k-iterable premouse, where $k < \omega$. Let r be the $k+1$st standard parameter of $(\mathcal{M}, u_k(\mathcal{M}))$. Then r is $k+1$-solid and $k+1$ universal over $(\mathcal{M}, u_k(\mathcal{M}))$.*

PROOF. Let $u = u_k(\mathcal{M})$ and $r = \langle \alpha_0, \cdots, \alpha_S \rangle$, with the ordinals α_s in decreasing order. Let $\alpha_{S+1} = \rho_{k+1}^{\mathcal{M}}$. Let $s \leq S+1$ be least such that

$$\mathrm{Th}_{k+1}^{\mathcal{M}}(\alpha_s \cup \{\alpha_0, \cdots, \alpha_{s-1}, u\}) \notin |\mathcal{M}|.$$

Such an s certainly exists, since $S+1$ will do. Let

$$\mathcal{H} = \mathcal{H}_{k+1}^{\mathcal{M}}(\alpha_s \cup \{\alpha_0, \cdots, \alpha_{s-1}, u\}),$$

let $\pi : \mathcal{H} \to \mathcal{M}$ be the inverse of the collapse (so that π is a k-embedding), and let $\bar{u} = \pi^{-1}(u)$ and $\bar{\alpha}_j = \pi^{-1}(\alpha_j)$ for $j < s$.

Our strategy is to compare $\mathcal{H}$ with $\mathcal{M}$, using k-maximal trees. Suppose that $\mathcal{P}$ is the model produced at the end on the $\mathcal{H}$ side, and $\mathcal{Q}$ the model produced on the $\mathcal{M}$ side. Suppose the branches $\mathcal{H}$ to $\mathcal{P}$ and $\mathcal{M}$ to $\mathcal{Q}$ involve no dropping of any kind, so that we have generalized $r\Sigma_{k+1}$ maps $i : \mathcal{H} \to \mathcal{P}$ and $j : \mathcal{M} \to \mathcal{Q}$. Suppose $\operatorname{crit} i \geq \alpha_s$ and $\operatorname{crit} j \geq \rho_{k+1}^{\mathcal{M}}$. Then

$$\mathrm{Th}_{k+1}^{\mathcal{H}}(\alpha_s \cup \{\bar{\alpha}_0, \cdots, \bar{\alpha}_{s-1}, \bar{u}\}) = \mathrm{Th}_{k+1}^{\mathcal{P}}(\alpha_s \cup \{i(\bar{\alpha}_0), \cdots, i(\bar{\alpha}_{s-1}), i(\bar{u})\}) \notin \mathcal{Q}$$

and

$$\mathrm{Th}^{\mathcal{M}}_{k+1}(\rho^{\mathcal{M}}_{k+1} \cup \{r, u\}) = \mathrm{Th}^{\mathcal{Q}}_{k+1}(\rho^{\mathcal{M}}_{k+1} \cup \{j(r), j(u)\}) \notin \mathcal{P}$$

so that neither of $\mathcal{P}$ and $\mathcal{Q}$ is a proper initial segment of the other, and hence $\mathcal{P} = \mathcal{Q}$.

Now if $\mathcal{M}$ is not $k+1$-solid then $s < S+1$ and hence $\rho^{\mathcal{M}}_{k+1} < \rho^{\mathcal{H}}_{k+1}$ because we didn't throw α_s as a member into the hull collapsing to $\mathcal{H}$. But we can show $\rho^{\mathcal{H}}_{k+1} \leq \rho^{\mathcal{P}}_{k+1}$, so $\rho^{\mathcal{Q}}_{k+1} \leq \rho^{\mathcal{M}}_{k+1} < \rho^{\mathcal{H}}_{k+1} \leq \rho^{\mathcal{P}}_{k+1}$ contradicting the fact that $\mathcal{P} = \mathcal{Q}$. Thus $\mathcal{M}$ is $k+1$-solid. It follows that $s = S+1$, and $\operatorname{crit} j \geq \rho^{\mathcal{M}}_{k+1}$ so we have $P^{\mathcal{M}}(\rho^{\mathcal{M}}_{k+1}) \subseteq |\mathcal{H}|$. Thus $\mathcal{M}$ is $k+1$-universal.

There are many problems in completing this sketch, but the main one is arranging that $\operatorname{crit} i \geq \alpha_s$. Our strategy will be to modify the comparison. Instead of comparing the models $\mathcal{M}$ and $\mathcal{H}$ by iteration trees $\mathcal{U}$ on $\mathcal{M}$ and $\mathcal{T}$ on $\mathcal{H}$, we will use a iteration tree $\mathcal{U}$ on the model $\mathcal{M}$ and a *pseudo-iteration tree* $\bar{\mathcal{T}}$ on the pair of models $(\mathcal{M}, \mathcal{H})$. The situation can be represented by the following diagram:

$$\begin{array}{ccccc} & & \mathcal{M}=\mathcal{P}_0 & \overset{\mathcal{T}}{\rule{6em}{0.4pt}} & \mathcal{P}_\theta \\ & \nearrow^{\pi_{-1}=\mathrm{id}} & \uparrow{\scriptstyle \pi_0} & & \uparrow{\scriptstyle \pi_\theta} \\ \mathcal{M}=\bar{\mathcal{P}}_{-1} & \rule{2em}{0.4pt} & \mathcal{H}=\bar{\mathcal{P}}_0 & \overset{\bar{\mathcal{T}}}{\rule{6em}{0.4pt}} & \bar{\mathcal{P}}_\theta \\ & & \mathcal{M}=\mathcal{Q}_0 & \overset{\mathcal{U}}{\rule{6em}{0.4pt}} & \mathcal{Q}_\theta \end{array}$$

The horizontal lines in this diagram indicate that the corresponding models are in the same tree, so that there is an embedding between them just in case they are on the same branch of the tree and there is no dropping on the branch between them. The comparison takes place between $\mathcal{U}$, which is a genuine iteration tree, and the pseudo-iteration tree $\bar{\mathcal{T}}$. The thing which makes $\bar{\mathcal{T}}$ a pseudo- iteration tree, rather than a real one, is that its underlying tree $\bar{T}$ has two separate roots, -1 and 0, corresponding to the models $\bar{\mathcal{P}}_{-1} = \mathcal{M}$ and $\bar{\mathcal{P}}_0 = \mathcal{H}$. We take $\rho_{-1} = \alpha_s$, and then we continue the comparison exactly as if $\bar{\mathcal{T}}$ were a real iteration tree. This means that whenever an extender $\bar{E}_\nu$ appears in the pseudo-tree $\bar{\mathcal{T}}$ such that $\operatorname{crit}(\bar{E}_\nu) < \alpha_s$ then $\bar{T}\text{-Pred}(\nu+1) = -1$, so that the $\nu+1$st model $\bar{\mathcal{P}}_{\nu+1}$ of $\bar{\mathcal{T}}$ is equal to $\mathrm{Ult}(\bar{\mathcal{P}}^*_\nu, E_\nu)$ for some initial segment $\mathcal{P}^*_\nu$ of $\mathcal{M}$.

Since $\bar{\mathcal{T}}$ is not a genuine iteration tree, we don't know directly that it has well founded branches. For this we use the iteration tree $\mathcal{T}$ and embeddings π_α, which are defined by setting $\pi_{-1} = \mathrm{id}$, letting π_0 be the inverse of the collapse map, and then using the shift lemma to copy $\bar{\mathcal{T}}$. Since $\mathcal{T}$ is a genuine iteration tree, theorem 6.2 implies that it is simple. Thus it has well founded branches at every stage, and the embeddings $\pi_\nu : \bar{\mathcal{P}}_\nu \to \mathcal{P}_\nu$ ensure that the corresponding branches of $\bar{\mathcal{T}}$ are also well founded.

We will show that $0\bar{T}\theta$ and that there is no dropping along the main branch of either tree. Thus the maps $\bar{\imath}_{0,\theta} : \mathcal{H} \to \bar{\mathcal{P}}^\theta$ and $i^{\mathcal{U}}_{0,\theta} : \mathcal{M} \to \mathcal{Q}_\theta$ are defined. In addition we show that $\bar{\mathcal{P}}_\theta = \mathcal{Q}_\theta$ and finally that $\bar{\imath}_{0,\theta} = i^{\mathcal{U}}_{0,\theta} \circ \pi_0$.

We now begin the actual proof of theorem 8.1. Notice first that if $\rho^{\mathcal{M}}_{k+1} > \operatorname{lh} E$ for all extenders E from the $\mathcal{M}$-sequence, then $\mathcal{H}$ is already an initial segment of $\mathcal{M}$ (since $\alpha_s \geq \rho^{\mathcal{M}}_{k+1}$). In this case, no iteration is necessary, and we have that $\mathcal{H} = \mathcal{M}$, which easily gives the desired results. Thus we may and do assume that $\rho^{\mathcal{M}}_{k+1} \leq \operatorname{lh} E$ for some extender E on the $\mathcal{M}$-sequence. According to the strong uniqueness theorem then, every k-maximal iteration tree on $\mathcal{M}$ is simple. This fact will make the Dodd-Jensen lemma applicable in what follows.

We now define by induction on length: (1) a "psuedo iteration tree" $\bar{\mathcal{T}}$ on the pair $(\mathcal{H}, \mathcal{M})$, (2) a tree $\mathcal{T}$ on $\mathcal{M}$ "enlarging" $\bar{\mathcal{T}}$, and (3) a tree $\mathcal{U}$ on $\mathcal{M}$. We use $\bar{\mathcal{P}}_\alpha$, $\mathcal{P}_\alpha$, and $\mathcal{Q}_\alpha$ for the αth models of $\bar{\mathcal{T}}$, $\mathcal{T}$, and $\mathcal{U}$ respectively. We use T for the tree ordering of $\mathcal{T}, \bar{T}$ for that of $\bar{\mathcal{T}}$, and U for the tree ordering of $\mathcal{U}$. The rest we indicate with superscripts; e.g., $\rho^{\mathcal{T}}_\alpha$, $\bar{\rho}_\alpha$, and $\rho^{\mathcal{U}}_\alpha$ or $i^{\mathcal{T}}_{\alpha\beta}$, $\bar{\imath}_{\alpha\beta}$, and $i^{\mathcal{U}}_{\alpha\beta}$.

The systems $\mathcal{T}$ and $\mathcal{U}$ will literally be a padded iteration trees on $\mathcal{M}$; they will be k-maximal and non-overlapping. $\bar{T}$ will not literally be a tree ordering in our sense, as it will have two roots, but will agree with T on $\mathrm{OR} - \{0, -1\}$.

Simultaneously with $\mathcal{T}$, $\bar{\mathcal{T}}$, and $\mathcal{U}$ we define $\pi_\alpha : \bar{\mathcal{P}}_\alpha \to \mathcal{P}_\alpha$ such that the map π_α is a weak $\deg(\alpha)$-embedding.

We begin by setting

$$\bar{\mathcal{P}}_{-1} = \mathcal{M},\ \bar{\mathcal{P}}_0 = \mathcal{H},\ \mathcal{P}_0 = \mathcal{M},\ \mathcal{Q}_0 = \mathcal{M}$$

and

$$\pi_{-1} = \text{identity},\ \pi_0 = \text{inverse of collapse}\,.$$

Notice that π_{-1} and π_0 are k-embeddings.

Now suppose that we have defined $\mathcal{T} \restriction \theta$, $\bar{\mathcal{T}} \restriction \theta$, and $\mathcal{U} \restriction \theta$. (This means we have defined the models $\bar{\mathcal{P}}_\alpha$, $\mathcal{P}_\alpha$, and $\mathcal{Q}_\alpha$ for $\alpha < \theta$, together with the extenders $\bar{E}_\alpha$, $E^{\mathcal{T}}_\alpha$, and $E^{\mathcal{U}}_\alpha$, for $\alpha + 1 < \theta$, etc.) Suppose we have also defined $\pi_\alpha : \bar{\mathcal{P}}_\alpha \to \mathcal{P}_\alpha$ for $\alpha < \theta$ with the following commutativity and agreement properties.

(i) If $\alpha \bar{T} \beta$ and $\bar{D} \cap (\alpha, \beta]_{\bar{T}} = \emptyset$ then $i^{\mathcal{T}}_{\alpha\beta} \circ \pi_\alpha = \pi_\beta \circ \bar{\imath}_{\alpha\beta}$.

(ii) If $0 \leq \alpha < \beta < \theta$, then $\bar{\mathcal{P}}_\alpha$ agrees with $\bar{\mathcal{P}}_\beta$ below $\operatorname{lh} \bar{E}_\alpha$; moreover letting $\gamma = \operatorname{lh} \bar{E}_\alpha$ and $N = J^{\bar{\mathcal{P}}_\alpha}_\gamma = J^{\bar{\mathcal{P}}_\beta}_\gamma$, we have $\pi_\alpha \restriction N = \pi_\beta \restriction N$.

Remark. Some simple observations about $\mathcal{H}$.

(1) We may assume $\alpha_s \in |\mathcal{H}|$. For otherwise $\mathcal{H}$ is an initial segment of $\mathcal{M}$ (if $\mathcal{M}$ and hence $\mathcal{H}$ is active, then the initial segment condition on good extender sequences implies $\dot{F}^{\mathcal{H}} = \dot{F}^{\mathcal{M}} \restriction \mathrm{OR}^{\mathcal{H}}$ is on the $\mathcal{M}$-sequence) but then $\mathcal{H} = \mathcal{M}$, and we are done.

(2) $\mathcal{H} \models \alpha_s$ is a cardinal, since $\alpha_s = \operatorname{crit} \pi_0$ if $s < S + 1$ and, $\alpha_s = \pi(\alpha_s) = \rho^{\mathcal{M}}_{k+1}$ if $s = S + 1$.

(3) For $\beta \geq 0$ and $\kappa < \alpha_s$, $P(\kappa) \cap |\bar{\mathcal{P}}_\beta| = P(\kappa) \cap |\mathcal{H}| = P(\kappa) \cap |\mathcal{J}^{\mathcal{M}}_{\alpha_s}|$. However, it seems possible at this point that $P(\kappa) \cap |\mathcal{M}|$ might be larger than $P(\kappa) \cap |\mathcal{H}|$.

We now define $\mathcal{T} \restriction \theta + 1$, $\bar{\mathcal{T}} \restriction \theta + 1$ and $\mathcal{U} \restriction \theta + 1$.

CASE 1. θ is a limit ordinal.

In this case, we have only to pick cofinal wellfounded branches in each of our trees.

As $\mathcal{T} \restriction \theta$ is k-maximal and $\rho^{\mathcal{M}}_{k+1} \leq \text{lh } E$ for some extender E from the $\mathcal{M}$ sequence, $\mathcal{T} \restriction \theta$ is simple. As $\mathcal{M}$ is k-iterable, there is a cofinal wellfounded branch b of $\mathcal{T}$. Similarly, there is a cofinal wellfounded branch c of $\mathcal{U}$. Finally, let $\bar{b} = b$ or $\bar{b} = (b - \{0\}) \cup \{-1\}$, whichever is a branch of $\bar{\mathcal{T}}$. Set

$$\begin{aligned} \mathcal{P}_\theta &= \text{direct limit of } \mathcal{P}_\alpha\,, \quad \alpha \in b - \sup D^{\mathcal{T}} \\ \bar{\mathcal{P}}_\theta &= \text{direct limit of } \bar{\mathcal{P}}_\alpha\,, \quad \alpha \in \bar{b} - \sup \bar{D} \\ \mathcal{Q}_\theta &= \text{direct limit of } \mathcal{Q}_\alpha\,, \quad \alpha \in c - \sup D^{\mathcal{U}} \end{aligned}$$

and extend $T, \bar{T}$ and U to $\theta + 1$ correspondingly. For $\alpha \in \bar{b} - \sup\ \bar{D}$ and $x \in |\bar{\mathcal{P}}_\alpha|$ we can set

$$\pi_\theta(\bar{\imath}_{\alpha,\theta}(x)) = i^{\mathcal{T}}_{\alpha,\theta}(\pi_\alpha(x))$$

(where of course $\bar{\imath}_{\alpha,\beta} = \bar{\imath}_{\alpha,\bar{b}}$, etc.), and by induction hypotheses (i) and (ii) this gives a well-defined $\pi_\theta : \bar{\mathcal{P}}_\theta \to \mathcal{P}_\theta$. Clearly π_θ is a $\overline{\deg}(\theta)$-embedding and (i) and (ii) continue to hold.

CASE 2. $\theta = \eta + 1$. In this case we "iterate the least disagreement" between $\bar{\mathcal{P}}_\eta$ and $\mathcal{Q}_\eta$, as in the proof of the comparison lemma.

Let γ be least $\leq \text{OR}^{\bar{\mathcal{P}}_\eta} \wedge \text{OR}^{\mathcal{Q}_\eta}$ such that

$$\mathcal{J}^{\bar{\mathcal{P}}_\eta}_\gamma \neq \mathcal{J}^{\mathcal{Q}_\eta}_\gamma\,;$$

if no such γ exists then we stop the construction of $\mathcal{T}$, $\bar{\mathcal{T}}$, and $\mathcal{U}$. Set

$$\bar{E}_\eta = \begin{cases} \dot{F}^{\mathcal{J}^{\bar{\mathcal{P}}_\eta}_\gamma}\,, & \text{if } \mathcal{J}^{\bar{\mathcal{P}}_\eta}_\gamma \text{ is active} \\ \varnothing & \text{otherwise} \end{cases}$$

$$E^{\mathcal{U}}_\eta = \begin{cases} \dot{F}^{\mathcal{J}^{\mathcal{Q}_\eta}_\gamma}\,, & \text{if } \mathcal{J}^{\mathcal{Q}_\eta}_\gamma \text{ is active} \\ \varnothing & \text{otherwise}\,. \end{cases}$$

On the $\mathcal{U}$ side the rest is determined by the demands of a k-maximal iteration tree. So $\mathcal{U}$-pred$(\eta + 1) = \xi$, where

$$\xi = \text{least } \alpha \text{ such that crit } E^{\mathcal{U}}_\eta < \rho^{\mathcal{U}}_\alpha\,.$$

(Assuming now $E_\eta^{\mathcal{U}} \neq \emptyset$; if $E_\eta^{\mathcal{U}} = \emptyset$ we just pad for one step.) Let $\kappa = \text{crit } E_\eta^{\mathcal{U}}$, let $Q_{\eta+1}^*$ be the longest initial segment $\mathcal{N}$ of Q_ξ such that $P(\kappa) \cap |\mathcal{N}| = P(\kappa) \cap |Q_\eta|$ and let

$$Q_{\eta+1} = \text{Ult}_n(Q_{\eta+1}^*, E_\eta^{\mathcal{U}})$$

where

$$n = \text{largest } s \text{ such that } \kappa < \rho_s^{Q_{\eta+1}^*} \text{ and } s \leq k \text{ if } D^{\mathcal{U}} \cap [0, \eta+1]_U = \emptyset\,.$$

On the $\bar{\mathcal{T}}$ side we proceed similarly. We assume $\bar{E}_\eta \neq \emptyset$; otherwise we pad for one step. Set $\kappa = \text{crit}(\bar{E}_\eta)$ and let let $\bar{\mathcal{T}}\text{-pred}(\eta+1) = \xi$, where

$$\xi = \text{least } \alpha \text{ such that } \kappa < \bar{\rho}_\xi\,,$$

so that in particular $\xi = -1$ if $\kappa < \alpha_i = \bar{\rho}_{-1}$. Now set $\bar{\mathcal{P}}_{\eta+1}^*$ equal to the longest initial segment $\mathcal{N}$ of $\bar{\mathcal{P}}_\xi$ such that $P(\kappa) \cap |\mathcal{N}| = P(\kappa) \cap |\bar{P}_\eta|$ and

$$\bar{\mathcal{P}}_{\eta+1} = \text{Ult}_n(\bar{\mathcal{P}}_{\eta+1}^*, \bar{E}_\eta)$$

where n is the largest integer s such that $\kappa < \rho_s^{Q_{\eta+1}^*}$ and such that $s \leq k$ if $\bar{D} \cap \{\alpha \mid \alpha \bar{T}(\eta+1) \vee \alpha = \eta+1\} = \emptyset$.

That $\bar{\mathcal{P}}_{\eta+1}$ agrees with $\bar{\mathcal{P}}_\eta$ below $\text{lh}\,\bar{E}_\eta$ is proved as usual. Notice that if $\xi = -1$, then as $P(\kappa) \cap |\mathcal{J}_{\alpha_i}^{\mathcal{M}}| = P(\kappa) \cap |\bar{\rho}_\eta|$, there is an $\mathcal{N}$ as called for in the definition of $\bar{\mathcal{P}}_{\eta+1}^*$.

Finally, we extend $\mathcal{T}$ by "copying" what we just did with $\bar{\mathcal{T}}$. Let γ be least such that $\mathcal{J}_\gamma^{\bar{\mathcal{P}}_\eta} \neq \mathcal{J}_\gamma^{Q_\eta}$. Assume that $\mathcal{J}_\gamma^{\bar{\mathcal{P}}_\eta}$ is active; otherwise we just pad $\mathcal{T}$ for one step. Let

$$E_\eta^{\mathcal{T}} = \dot{F}^{\mathcal{N}}, \quad \text{where} \quad \mathcal{N} = \mathcal{J}_{\pi_\eta(\gamma)}^{\mathcal{P}_\eta},$$

where as usual we let $\pi_\eta(\text{OR}^{\bar{\mathcal{P}}_\eta}) = \text{OR}^{\mathcal{P}_\eta}$.

Subcase A. $\bar{\mathcal{T}}\text{-pred}(\eta+1) = -1$.

Let $\mathcal{T}\text{-pred}(\eta+1) = 0$,

$$\mathcal{P}_{\eta+1}^* = \bar{\mathcal{P}}_{\eta+1}^*$$
$$\mathcal{P}_{\eta+1} = \text{Ult}_n(\mathcal{P}_{\eta+1}^*, E_\eta^{\mathcal{T}}), \quad \text{where } n = \overline{\deg}\ (\eta+1)\,.$$

We get $\pi_{\eta+1} : \bar{\mathcal{P}}_{\eta+1} \to \mathcal{P}_{\eta+1}$ by the shift lemma, lemma 5.2 which implies that $\pi_{\eta+1}$ is a $\overline{\deg}(\eta+1)$-embedding with the required commutativity and agreement properties (i) and (ii).

Subcase B. $\bar{\mathcal{T}}\text{-pred}(\eta+1) = \xi \geq 0$. Let $\mathcal{T}\text{-pred}(\eta+1) = \xi$. Let $\bar{\mathcal{P}}_{\eta+1}^* = \mathcal{J}_\nu^{\bar{\mathcal{P}}_\xi}$; then

$$\mathcal{P}_{\eta+1}^* = \mathcal{J}_{\pi_\xi(\nu)}^{\mathcal{P}_\xi}$$

where $\pi_\xi(\mathrm{OR}^{\bar{\mathcal{P}}_\xi}) = \mathrm{OR}^{\mathcal{P}_\xi}$. Let $n = \overline{\deg}(\eta+1)$, then

$$\mathcal{P}_{\eta+1} = \mathrm{Ult}_n(\mathcal{P}^*_{\eta+1}, E^{\mathcal{T}}_\eta).$$

Finally, we get the desired $\pi_{\eta+1}$ by the shift lemma.

This completes the construction of $\mathcal{T}$, $\bar{\mathcal{T}}$, and $\mathcal{U}$. We leave it to the reader to check the many details we ought to have verified in the course of the construction. (In particular, that $\mathcal{T}$ is a k-maximal iteration tree, and that the π_η's have the required commutativity and agreement properties.)

Now because $\mathcal{T}$ and $\mathcal{U}$ are simple we must reach an ordinal θ such that $\bar{\mathcal{P}}_\theta$ is an initial segment of $\mathcal{Q}_\theta$ or vice-versa. The proof is exactly the same as the proof in §7 that the comparison process stops.

We shall say that a branch b of $\mathcal{U}$ *drops* if either $D^{\mathcal{U}} \cap b \neq \emptyset$ or $\exists \alpha \in b\ (\overline{\deg}^{\mathcal{U}}(\alpha) \neq k)$, and similarly for branches of $\bar{\mathcal{T}}$ or $\mathcal{T}$.

We need to verify that, just as with the comparison in section 7, at most one side of the comparison drops, and that the side which drops is the longer. That is, if the main branch $\{\beta : \beta \bar{T} \theta\}$ of $\bar{\mathcal{T}}$ drops then the main branch $[0,\theta]_U$ of $\mathcal{U}$ does not drop and $\mathcal{P}_\theta$ is not a proper initial segment of $\mathcal{Q}_\theta$, while if the main branch of $\mathcal{U}$ drops then the main branch of $\mathcal{T}$ does not drop and $\mathcal{Q}_\theta$ is not a proper initial segment of $\bar{\mathcal{P}}_\theta$.

It is immediate that if either branch drops then its final model is not ω-sound, and hence cannot be a proper initial segment of the final model of the other branch. If follows that if both branches dropped then we would have $\bar{\mathcal{P}}_\theta = \mathcal{Q}_\theta$. This implies that if the last drop on $\{\beta : \beta \bar{T} \theta\}$ occurs at $\alpha+1$ and the last drop on $[0,\theta]_u$ at $\beta+1$, then

$$\begin{aligned} \deg^{\mathcal{U}}(\beta+1) &= \text{the least } n \text{ such that } \mathcal{Q}_\theta \text{ is not } n+1 \text{ sound} \\ &= \text{the least } n \text{ such that } \bar{\mathcal{P}}_\theta \text{ is not } n+1 \text{ sound} \\ &= \overline{\deg}(\alpha+1). \end{aligned}$$

Moreover, if $n = \deg^{\mathcal{U}}(\beta+1)$,

$$\begin{aligned} \mathcal{Q}^*_{\beta+1} &= \mathfrak{C}_{n+1}(\mathcal{Q}^*_{\beta+1}) = \mathfrak{C}_{n+1}(\mathcal{Q}_\theta) \\ &= \mathfrak{C}_{n+1}(\bar{\mathcal{P}}_\theta) = \mathfrak{C}_{n+1}(\bar{\mathcal{P}}^*_{\alpha+1}) = \bar{\mathcal{P}}^*_{\alpha+1}. \end{aligned}$$

Also

$$\mathrm{crit}\ i^{\mathcal{U}}_{\beta+1,\theta} \circ i^{*\mathcal{U}}_{\beta+1} \geq \rho^{\mathcal{Q}^*_{\beta+1}}_{n+1} = \rho^{\mathcal{Q}_\theta}_{n+1} = \rho^{\bar{\mathcal{P}}_\theta}_{n+1},$$

and

$$\mathrm{crit}\ \bar{\imath}_{\alpha+1,\theta} \circ \bar{\imath}^*_{\alpha+1} \geq \rho^{\bar{\mathcal{P}}^*_{\alpha+1}}_{n+1} = \rho^{\bar{\mathcal{P}}_\theta}_{n+1}.$$

Also

$$\begin{aligned}\text{crit } E^{\mathcal{U}}_\beta &= \text{least } \kappa \text{ such that } \kappa \neq \tau^{Q_\theta}[\bar{\alpha}, p_{n+1}(Q_\theta)] \text{ for any} \\ &\qquad \tau \in \text{Sk}_{n+1} \text{ and } \bar{\alpha} \in (\rho_{n+1}^{Q_\theta})^{<\omega} \\ &= \text{least } \kappa \text{ such that } \kappa \neq \tau^{\bar{\mathcal{P}}_\theta}[\bar{\alpha}, p_{n+1}(\bar{\mathcal{P}}_\theta)] \text{ for any} \\ &\qquad \tau \in \text{Sk}_{n+1} \text{ and } \bar{\alpha} \in (\rho_{n+1}^{Q_\theta})^{<\omega} \\ &= \text{crit } \bar{E}_\alpha .\end{aligned}$$

Finally, if $A \in Q^*_{\beta+1}$ and $A \subseteq [\text{crit } E^{\mathcal{N}}_\beta]^n$, then letting

$$A = \tau^{Q^*_{\beta+1}}[\bar{\alpha}, p_{n+1}(Q^*_{\beta+1})]$$

where $\bar{\alpha} \in (\rho_{n+1}^{Q^*_{\beta+1}})^{<\omega}$,

$$\begin{aligned} i^{\mathcal{U}}_{\beta+1,\theta} \circ i^{*\mathcal{U}}_{\beta+1}(A) &= \tau^{Q_\theta}[\bar{\alpha}, p_{n+1}(Q_\theta)] \\ &= \tau^{\bar{\mathcal{P}}_\theta}[\bar{\alpha}, p_{n+1}(\bar{\mathcal{P}}_\theta)] \\ &= \bar{\imath}_{\alpha+1,\theta} \circ \bar{\imath}^*_{\alpha+1}(A) . \end{aligned}$$

It follows that one of $E^{\mathcal{U}}_\beta$ and $\bar{E}_\alpha$ is an initial segment of the other, and this is a contradiction as in the proof of the comparison lemma. Thus at most one of the trees $\bar{\mathcal{T}}$ and $\mathcal{U}$ can have a drop along its main branch.

CLAIM 1. $0\bar{T}\theta$, that is, $\bar{\mathcal{P}}_\theta$ lies above $\mathcal{P}_0 = \mathcal{H}$ in the $\bar{\mathcal{T}}$ system.

PROOF. Assume not; that is, assume that $-1\bar{T}\theta$, so that $\bar{\mathcal{P}}_\theta$ lies above $\bar{\mathcal{P}}_{-1} = \mathcal{M}$. We know that at least one of the branches $[-1,\theta]_{\bar{T}}$ and $[0,\theta]_{\mathcal{U}}$ does not drop.

CASE 1. $[-1,\theta]_{\bar{T}}$ drops.

Then $\bar{\mathcal{P}}_\theta$ is not ω-sound, so is not a proper initial segment of Q_θ. Suppose first Q_θ is a proper initial segment of $\bar{\mathcal{P}}_\theta$; say $Q_\theta = \mathcal{J}^{\bar{\mathcal{P}}_\theta}_\gamma$. Let $\sigma = \pi_\theta \restriction \mathcal{J}^{\bar{\mathcal{P}}_\theta}_\gamma$, so that $\sigma : Q_\theta \to \mathcal{J}^{\mathcal{P}_\theta}_{\pi_\theta(\gamma)}$ is fully elementary. Then the map $\sigma \circ i^{\mathcal{U}}_{0,\theta}$ is a weak k-embedding from $\mathcal{M}$ to a proper initial segment of $\mathcal{P}_\theta$. As $\mathcal{T}$ is k-bounded and simple, this contradicts the Dodd-Jensen lemma.

Suppose next that $Q_\theta = \bar{\mathcal{P}}_\theta$. Then as Q_θ is k-sound, $\overline{\deg}(\theta) \geq k$, so that π_θ is a weak k-embedding. Thus $\pi_\theta \circ i^{\mathcal{U}}_{0,\theta}$ is a weak k-embedding from $\mathcal{M}$ to $\mathcal{P}_\theta$. But by case hypothesis, $[0,\theta]_T$ drops. This contradicts the Dodd-Jensen lemma. (As $\deg^{\mathcal{T}}(\theta) \geq k$, we must have $D^{\mathcal{T}} \cap [0,\theta]_T \neq \varnothing$.)

CASE 2. $[0,\theta]_U$ drops.

In this case, $[-1,\theta]_{\bar{T}}$ doesn't drop and $\bar{\mathcal{P}}_\theta$ is an initial segment of Q_θ. If proper, then $\bar{\imath}_{-1,\theta}$ is a k-embedding of $\mathcal{M}$ to a proper initial segment of Q_θ, which lies on a k-bounded, simple iteration tree with base model $\mathcal{M}$. This contradicts Dodd-Jensen. If $\bar{\mathcal{P}}_\theta = Q_\theta$, then as $\bar{\mathcal{P}}_\theta$ is k-sound, $\deg^{\mathcal{U}}(\theta) \geq k$. But then the fact that $[0,\theta]_U$ drops contradicts Dodd-Jensen.

CASE 3. Neither $[-1,\theta]_{\bar{T}}$ nor $[0,\theta]_U$ drops.

If $\bar{\mathcal{P}}_\theta$ is a proper initial segment of Q_θ, then $\bar{\imath}_{-1,\theta}$ contradicts Dodd-Jensen. If Q_θ is proper initial segment of $\bar{\mathcal{P}}_\theta$, then $\pi_\theta \circ i^{\mathcal{U}}_{0,\theta}$ contradicts Dodd-Jensen. So we must have $\bar{\mathcal{P}}_\theta = Q_\theta$. We now use the minimality of the iteration maps $i^{\mathcal{U}}_{0,\theta}$, $i^{\mathcal{T}}_{0,\theta}$ given by the Dodd-Jensen lemma. Let $\leq_{\text{L}}$ be the order of construction in premice.

Fix any $x \in |\mathcal{M}|$. By Dodd-Jensen,

$$i^{\mathcal{U}}_{0,\theta}(x) \leq_{\text{L}} \bar{\imath}_{-1,\theta}(x)$$

since $i^{\mathcal{U}}_{0,\theta}$ is a "k-bounded" iteration map, and $\bar{\imath}_{-1,\theta}$ a k-embedding. But also

$$i^{\mathcal{T}}_{0,\theta}(x) \leq_{\text{L}} \pi_\theta(i^{\mathcal{U}}_{0,\theta}(x))$$

since $i^{\mathcal{T}}_{0,\theta}$ is a k-bounded iteration map and $i^{\mathcal{U}}_{0,\theta} \circ \pi_\theta$ is a weak k-embedding. Then

$$\pi_\theta(\bar{\imath}_{-1,\theta}(x)) = i^{\mathcal{T}}_{0,\theta}(x) \leq_{\text{L}} \pi_\theta(i^{\mathcal{U}}_{0,\theta}(x))$$

so

$$\bar{\imath}_{-1,\theta}(x) \leq_{\text{L}} i^{\mathcal{U}}_{0,\theta}(x)$$

so

$$\bar{\imath}_{-1,\theta}(x) = i^{\mathcal{U}}_{0,\theta}(x)\,.$$

But if $\bar{\imath}_{-1,\theta} = i^{\mathcal{U}}_{0,\theta}$, then the first extender used on $[-1,\theta]_{\bar{T}}$ is compatible with the first extender used on $[0,\theta]_U$, which is impossible. □

This proves Claim 1, and it follows that $[0,\theta]_{\bar{T}}$ is the main branch of $\bar{\mathcal{T}}$. Again, we know that at most one of the branches $[0,\theta]_{\bar{T}}$ and $[0,\theta]_U$ drops.

CLAIM 2. $[0,\theta]_{\bar{T}}$ doesn't drop.

PROOF. Suppose it did drop. Then $[0,\theta]_U$ does not drop and $\bar{\mathcal{P}}_\theta$ is not a proper initial segment of Q_θ. Suppose Q_θ is a proper initial segment of $\bar{\mathcal{P}}_\theta$. Then $\pi_\theta \circ i^{\mathcal{U}}_{0,\theta}$ is a weak k-embedding from $\mathcal{M}$ to a proper initial segment of $\mathcal{P}_\theta$, contrary to Dodd-Jensen. Suppose $Q_\theta = \bar{\mathcal{P}}_\theta$. Then as Q_θ is k-sound, $\overline{\deg}\,(\theta) \geq k$, so that $\pi_\theta \circ i^{\mathcal{U}}_{0,\theta}$ is a weak k-embedding from $\mathcal{M}$ to $\mathcal{P}_\theta$. As $[0,\theta]_T$ drops and $\overline{\deg}^{\mathcal{T}}\,(\theta) \geq k$, we have $D^{\mathcal{T}} \cap [0,\theta]_T \neq \varnothing$. This contradicts Dodd-Jensen. □

CLAIM 3. Q_θ is an initial segment of $\bar{\mathcal{P}}_\theta$.

PROOF. Claims 1 and 2 together imply that

$$\text{Th}^{\mathcal{H}}_{k+1}(\alpha_s \cup \{\langle \bar{\alpha}_0, \cdots, \bar{\alpha}_{s-1}, \bar{u}\rangle\}) =$$
$$\text{Th}^{\bar{\mathcal{P}}_\theta}_{k+1}(\alpha_s \cup \{\bar{\imath}_{0,\theta}(\langle \bar{\alpha}_0, \cdots, \bar{\alpha}_{s-1}\bar{u}\rangle)\}) \notin |\bar{\mathcal{P}}_\theta|\,.$$

Moreover, as $\mathrm{Th}_{k+1}^{\mathcal{H}}(\alpha_s \cup \{\langle\bar{\alpha}_0, \cdots, \bar{\alpha}_{s-1}, \bar{u}\rangle\})$ is essentially a subset of α_s, and is not in $\mathcal{M}$, it is not in Q_θ. (Note here that $P(\alpha_s)^{Q_1} \subseteq |Q_0|$, and that if $\xi \geq 1$ and $E_\xi^{\mathcal{U}} \neq \emptyset$, then $\mathrm{lh}\, E_\xi^{\mathcal{U}} > \alpha_s$, so that $P(\alpha_s) \cap |Q_{\xi+1}| \subseteq P(\alpha_s) \cap |Q_\xi|$, with equality holding after the least such ξ.) It follows that $\bar{\mathcal{P}}_\theta$, over which the subset of α_s in question is definable, is not a proper initial segment of Q_θ. □

CLAIM 4. $[0, \theta]_U$ doesn't drop.

Suppose otherwise. Then Q_θ is not ω-sound, so $Q_\theta = \bar{\mathcal{P}}_\theta$. But then Q_θ is k-sound, so that $\deg^{\mathcal{U}}(\theta) \geq k$.

Let $\gamma + 1$ be the largest member of $D^{\mathcal{U}} \cap [0, \theta]_{\mathcal{U}}$. Thus $\deg^{\mathcal{U}}(\xi) \geq k$ for all $\xi \geq \gamma + 1$ such that $\xi \in [0, \theta]_U$.

For any $X \subseteq |\mathcal{N}|$, any j, let

$$\overline{\mathrm{Th}}_j^{\mathcal{N}}(X) = \{(\varphi, \bar{a}) \in \mathrm{Th}_j^{\mathcal{N}}(X) \mid \varphi \quad \text{is pure } r\Sigma_j\}.$$

Then set

$$A = \overline{\mathrm{Th}}_{k+1}^{\bar{\mathcal{P}}_\theta}(\alpha_s \cup \{\bar{\imath}_{0,\theta}(\langle\bar{\alpha}_0, \cdots, \bar{\alpha}_{s-1}, \bar{u}\rangle)\}).$$

Thus A is $r\Sigma_{k+1}^{\bar{\mathcal{P}}_\theta}$, and by Lemma 2.10, $A \notin |\bar{\mathcal{P}}_\theta|$.

CASE 1. $\mathrm{crit}(i_{\gamma+1,\theta}^{\mathcal{U}} \circ i_{\gamma+1}^{*\mathcal{U}}) \geq \alpha_s$.

As in the proof of Lemma 4.5, we can show by induction on $\beta \in [\gamma + 1, \theta]_U$ that any set $X \subseteq \alpha_s$ which is $r\Sigma_{k+1}^{Q_\beta}$ is in fact $r\Sigma_{k+1}^{Q_{\gamma+1}^*}$. Thus A is $r\Sigma_{k+1}^{Q_{\gamma+1}^*}$. Thus $A \in Q_\xi$, where $\xi = U\text{-pred}(\gamma+1)$. But then $A \in |\mathcal{M}| = |Q_0|$, since $A \subseteq \alpha_s$. But then the proof of 2.10 shows that

$$\mathrm{Th}_{k+1}^{\bar{\mathcal{P}}_\theta}(\alpha_s \cup \{\bar{\imath}_{0,\theta}(\langle\bar{\alpha}_0, \cdots, \bar{\alpha}_{s-1}, \bar{u}\rangle)\}) \in |\mathcal{M}|,$$

a contradiction.

CASE 2. Otherwise. Let

$$\kappa = \mathrm{crit}(i_{\gamma+1,\theta}^{\mathcal{U}} \circ i_{\gamma+1}^{*\mathcal{U}}) < \alpha_s.$$

Since $\kappa = \mathrm{crit}(i_{\gamma+1}^{*\mathcal{U}}) = \mathrm{crit}\, E_\gamma^{\mathcal{U}}$, and $\gamma + 1 \in D^{\mathcal{U}}$, we have

$$P(\kappa) \cap |Q_{\gamma+1}^*| \not\subseteq P(\kappa) \cap |\mathcal{M}|.$$

Let ξ be least such that $E_\xi^{\mathcal{U}} \neq \emptyset$; thus $\xi \leq \gamma$ and $Q_\xi = Q_0 = \mathcal{M}$. Now $\mathcal{M}$ agrees with $Q_{\gamma+1}^*$ below $\mathrm{lh}\, E_\xi^{\mathcal{U}}$, and $\mathrm{lh}\, E_\xi^{\mathcal{U}} \geq \alpha_s$; thus there must be a subset of κ in $\mathcal{M}$ but not in $\mathcal{J}_{\alpha_s}^{\mathcal{M}}$. So

$$\mathcal{M} \models \mathrm{card}(\alpha_s) \leq \kappa.$$

Thus $\alpha_s \neq \rho_{k+1}^{\mathcal{M}}$ and $s < S+1$ and $\alpha_s = \text{crit } \pi_0$ where $\pi_0 : \mathcal{H} \to \mathcal{M}$ is the inverse of the collapse. But then

$$\mathcal{H} \models \alpha_s = \kappa^+ .$$

Thus $P(\kappa) \cap |\mathcal{H}| = P(\kappa) \cap J_{\alpha_s}^{\mathcal{H}} = P(\kappa) \cap |\bar{\mathcal{P}}_\eta|$, all $\eta \geq 0$. Now since Q_ξ agrees with $\bar{\mathcal{P}}_\xi$ below $\text{lh } E_\xi^{\mathcal{U}}$, and $\text{lh } E_\xi^{\mathcal{U}}$ is a cardinal of Q_η for $\eta > \xi$, we have

$$P(\kappa) \cap |Q_\eta| = P(\kappa) \cap J_{\alpha_s}^{Q_\eta} \qquad (\text{all } \eta > \xi),$$

and

$$Q_\eta \models \alpha_s = \kappa^+ \qquad (\text{all } \eta > \xi).$$

It follows, since $\gamma + 1 \in D^{\mathcal{U}}$, that $U\text{-pred}(\gamma+1) = \xi$. Also,

$$\text{crit}\,(i_{\gamma+1,\theta}^{\mathcal{U}}) \geq \rho_\gamma^{\mathcal{U}} \geq (\kappa^+)^{J_{\text{lh } E_\gamma^{\mathcal{U}}}^{Q_\gamma}} \geq \alpha_s .$$

We can then show by an induction using the proof of 4.5 that

$$A \in r\Sigma_{k+1}^{Q_{\gamma+1}} .$$

Say A is $r\Sigma_{k+1}^{Q_{\gamma+1}}$ in the parameter p, where $p = [a, f]_{E_\gamma^{\mathcal{U}}}^{Q_{\gamma+1}^*}$. It will be enough to show that $E_\gamma^{\mathcal{U}} \restriction \alpha_s \cup a$ is a member of $\mathcal{M}$, for then, since $Q_{\gamma+1}^* \in Q_\xi = \mathcal{M}$, we get that $A \in |\mathcal{M}|$, a contradiction.

Suppose first $\gamma = \xi$. Since $\gamma + 1 \in D^{\mathcal{U}}$, $E_\gamma^{\mathcal{U}} \neq \dot{F}^{Q_\xi}$, and thus $E_\gamma^{\mathcal{U}} \in Q_\xi$, as desired.

Now let $\gamma > \xi$. Since $\xi U \eta$ for all $\eta > \xi$, $\xi U \gamma$. If $E_\gamma^{\mathcal{U}} \neq \dot{F}^{Q_\gamma}$, then $E_\gamma^{\mathcal{U}} \in |Q_\gamma|$, and since $E_\gamma^{\mathcal{U}} \restriction \alpha_s \cup a$ is a subset of α_s, $E_\gamma^{\mathcal{U}} \restriction \alpha_s \cup a \in |Q_\xi|$, as desired. So we may assume that $E_\gamma^{\mathcal{U}} = \dot{F}^{Q_\gamma}$.

Now $D^{\mathcal{U}} \cap [\xi, \gamma]_{\mathcal{U}} \neq \emptyset$, as otherwise since $\text{crit } \dot{F}^{Q_\gamma} = \kappa$, $\text{crit } i_{\xi,\gamma}^{\mathcal{U}} > \kappa$ and $P(\kappa) \cap |Q_\gamma| = P(\kappa) \cap |Q_\xi|$.

So let $\eta + 1$ be largest in $D^{\mathcal{U}} \cap [\xi, \gamma]_{\mathcal{U}}$. So $\dot{F}^{Q_{\eta+1}^*}$ has critical point κ, and $i_{\eta+1,\gamma}^{\mathcal{U}} \circ i_{\eta+1}^*$ has critical point $> \kappa$, hence $\geq \alpha_s$. But now $\dot{F}^{Q_\gamma} \restriction (\alpha_s \cup a)$ is an $r\Sigma_1^{Q_\gamma}$ subset of α_s, and hence (as in the proof of 4.5) $\dot{F}^{Q_\gamma} \restriction \alpha_s \cup a$ is $r\Sigma_1^{Q_{\eta+1}^*}$. Since $\eta + 1 \in D^{\mathcal{U}}$, we get $\dot{F}^{Q_\gamma} \restriction (\alpha_s \cup a) \in Q_\xi$, as desired. □

CLAIM 5. $\bar{\mathcal{P}}_\theta = Q_\theta$.

PROOF. Otherwise Q_θ is a proper initial segment of $\bar{\mathcal{P}}_\theta$. But then $\pi_\theta \circ i_{0,\theta}^{\mathcal{U}}$ is a weak k-embedding from $\mathcal{M}$ to a proper initial segment of $\mathcal{P}_\theta$, which is on a k-bounded simple iteration tree based on $\mathcal{M}$. This contradicts Dodd-Jensen. □

CLAIM 6. $\bar{\imath}_{0,\theta}(\bar{u}) = i_{0,\theta}(u)$, and for $j \leq s-1$, $\bar{\imath}_{0,\theta}(\bar{\alpha}_j) = i^{\mathcal{U}}_{0,\theta}(\alpha_j)$.

PROOF. $\bar{\imath}_{0,\theta}(\bar{u}) = u_k(\bar{\mathcal{P}}_\theta) = u_k(Q_\theta) = i^{\mathcal{U}}_{0,\theta}(u)$, since $\bar{\imath}_{0,\theta}$ and $i_{0,\theta}(u)$ are k-embeddings.

We show the second assertion by induction on j. Assume it for $p < j$. As $i^{\mathcal{U}}_{0,\theta}$ is a k-embedding, the proofs of 4.6 and 4.7 show that

$$\mathrm{Th}^{Q_\theta}_{k+1}(i^{\mathcal{U}}_{0,\theta}(\alpha_j) \cup \{i^{\mathcal{U}}_{0,\theta}(\langle \alpha_0, \cdots, \alpha_{j-1}, u\rangle)\}) \in |Q_\theta|.$$

On the other hand

$$\mathrm{Th}^{\bar{\mathcal{P}}_\theta}_{k+1}(\bar{\imath}_{0,\theta}(\bar{\alpha}_j + 1) \cup \{\bar{\imath}_{0,\theta}(\langle \bar{\alpha}_0, \cdots, \bar{\alpha}_{j-1}, \bar{u}\rangle)\} \notin \bar{\mathcal{P}}_\theta.$$

So our induction hypothesis implies that $i^{\mathcal{U}}_{0,\theta}(\alpha_j) \leq \bar{\imath}_{0,\theta}(\bar{\alpha}_j)$. On the other hand, since the iteration map $i^{\mathcal{T}}_{0,\theta}$ is minimal and $\pi_\theta \circ i^{\mathcal{U}}_{0,\theta}$ is a k-embedding of $\mathcal{M}$ into $\mathcal{P}_\theta$, we have

$$i^{\mathcal{T}}_{0,\theta}(\alpha_j) \leq \pi_\theta(i^{\mathcal{U}}_{0,\theta}(\alpha_j))$$

or

$$\pi_\theta(\bar{\imath}_{0,\theta}(\bar{\alpha}_j)) \leq \pi_\theta(i^{\mathcal{U}}_{0,\theta}(\alpha_j))$$

so that $\bar{\imath}_{0,\theta}(\bar{\alpha}_j) \leq i^{\mathcal{U}}_{0,\theta}(\alpha_j)$, and thus $\bar{\imath}_{0,\theta}(\bar{\alpha}_j) = i^{\mathcal{U}}_{0,\theta}(\alpha_j)$, as desired. □

CLAIM 7. crit $i^{\mathcal{U}}_{0,\theta} \geq \rho^{\mathcal{M}}_{k+1}$.

PROOF. Assume not, and let $\kappa =$ crit $i^{\mathcal{U}}_{0,\theta} =$ crit $E^{\mathcal{U}}_\beta$, where $\beta + 1 \in [0, \theta]_{\mathcal{U}}$ is such that U-pred$(\beta + 1) = 0$. Then

$$\mathrm{Th}^{\mathcal{M}}_{k+1}(\kappa \cup \{\langle \alpha_0, \cdots, \alpha_{s-1}, u\rangle\}) \in |\mathcal{M}|.$$

It follows as in the proof of 4.6 that

$$\mathrm{Th}^{Q_{\beta+1}}_{k+1}(i_{0,\beta+1}(\kappa) \cup \{i_{0,\beta+1}(\langle \alpha_0, \cdots, \alpha_{s-1}, u\rangle)\}) \in |Q_{\beta+1}|.$$

But now $\alpha_s \leq \mathrm{lh}\, E^{\mathcal{U}}_\beta < i_{0,\beta+1}(\kappa)$, so

$$\mathrm{Th}^{Q_{\beta+1}}_{k+1}(\alpha_s \cup \{i^{\mathcal{U}}_{0,\beta+1}(\langle \alpha_0, \cdots, \alpha_{s-1}, u\rangle)\}) \in |Q_{\beta+1}|.$$

So, again using the proof of 4.6 if crit $i^{\mathcal{U}}_{\beta+1,\theta} < \alpha_s$ (which seems possible; we may have $\rho^{\mathcal{U}}_\beta < \alpha_s$),

$$\mathrm{Th}^{Q_\theta}_{k+1}(\alpha_s \cup \{i^{\mathcal{U}}_{0,\theta}(\langle \alpha_0, \cdots, \alpha_{s-1}, u\rangle)\}) \in |Q_\theta|.$$

This contradicts the conjunction of our previous claims. □

CLAIM 8. $s = S + 1$; that is, $\langle \alpha_0, \cdots, \alpha_S\rangle$ is $k + 1$-solid over $(\mathcal{M}, u)$.

PROOF. Let $A \subseteq \rho_{k+1}^{\mathcal{M}}$ be $r\Sigma_{k+1}^{\mathcal{M}}$ but not a member of $|\mathcal{M}|$. Then A is $r\Sigma_{k+1}^{Q_\theta}$, hence $r\Sigma_{k+1}^{\bar{\mathcal{P}}_\theta}$, hence $r\Sigma_{k+1}^{\mathcal{H}}$. But if $s < S+1$, this means A is $r\Sigma_{k+1}^{\mathcal{M}}$ in a parameter from $(\alpha_s \cup \{\alpha_0, \cdots, \alpha_{s-1}, u\})^{<\omega}$, hence in u and a parameter $<_{\text{lex}}$ $\langle \alpha_0, \cdots, \alpha_S \rangle$. This contradicts the minimality of $\langle \alpha_0, \cdots, \alpha_S \rangle$. □

CLAIM 9. $\mathcal{P}(\rho_{k+1}^{\mathcal{M}})^{\mathcal{M}} = \mathcal{P}(\rho_{k+1}^{\mathcal{M}})^{\mathcal{H}}$; that is, r is $k+1$-universal over $(\mathcal{M}, u)$.

PROOF. This follows easily from the facts that $\bar{\mathcal{P}}_\theta = Q_\theta$ and crit $\bar{\imath}_{0,\theta} \geq \rho_{k+1}^{\mathcal{M}}$, crit $i_{0,\theta}^{\mathcal{U}} \geq \rho_{k+1}$. □

This completes the proof of Theorem 8.1. □

The method by which 8.1 was proved gives some condensation results for 1-small coremice. One which will be of use to us is the following.

Theorem 8.2. *Let $\mathcal{H}$ and $\mathcal{M}$ be 1-small coremice, and suppose there is a non-trivial fully elementary $\pi : \mathcal{H} \to \mathcal{M}$ such that crit $(\pi) = \rho_\omega^{\mathcal{H}}$. Then either*

(a) *$\mathcal{H}$ is a proper initial segment of $\mathcal{M}$*

or

(b) *There is an extender E on the $\mathcal{M}$ sequence such that* $\operatorname{lh} E = \rho_\omega^{\mathcal{H}}$ *and $\mathcal{H}$ is a proper initial segment of* $\mathrm{Ult}_0(\mathcal{M}, E)$.

Remark. In case (b), H is not an initial segment of $\mathcal{M}$. The following example shows that case (b) can occur. Suppose $\mathcal{P}$ is an active 1-small coremouse, $\kappa =$ crit $\dot{F}^{\mathcal{P}}$, and $\dot{F}^{\mathcal{P}} \restriction \alpha$ is on the $\mathcal{P}$ sequence for some $\alpha > (\kappa^+)\mathcal{P}$. (We shall later construct such a $\mathcal{P}$.) Let

$$\sigma : \mathrm{Ult}_0(\mathcal{P}, \dot{F}^{\mathcal{P}} \restriction \alpha) \to \mathrm{Ult}_0(\mathcal{P}, \dot{F}^{\mathcal{P}})$$

be the natural embedding. It is easy to see $\alpha = $ crit (σ). Let

$$\mathcal{H} = (J_{\alpha+1}^{\dot{E}^{\mathcal{P}} \restriction \alpha}, \in, \dot{E}^{\mathcal{P}} \restriction \alpha)$$

and

$$\mathcal{M} = \sigma(\mathcal{H}), \quad \pi = \sigma \restriction \mathcal{H}.$$

Clearly $\alpha = \mathrm{crit}(\pi) = \rho_\omega^{\mathcal{H}}$, π is fully elementary, and $\mathcal{H}$ is not an initial segment of $\mathcal{M}$.

PROOF OF 8.2. Suppose first that $\operatorname{lh} E < \rho_\omega^{\mathcal{H}}$ for all extenders E from the $\mathcal{H}$ sequence. Then either $\mathcal{H}$ is an initial segment of $\mathcal{M}$, so that (a) holds, or we have a first E from the $\mathcal{M}$ sequence such that $\rho_\omega^{\mathcal{H}} \leq \operatorname{lh} E \leq \mathrm{OR}^{\mathcal{H}}$. As $\mathcal{M}$ is internally iterable, $\operatorname{lh} E$ is a cardinal of $L[\dot{E}^{\mathcal{M}} \restriction \rho_\omega^{\mathcal{H}}]$. But $\mathrm{card}(\mathrm{OR}^{\mathcal{H}}) \leq \rho_\omega^{\mathcal{H}}$ in $L[\dot{E}^{\mathcal{M}} \restriction \rho_\omega^{\mathcal{H}}]$, so $\operatorname{lh} E = \rho_\omega^{\mathcal{H}}$. Moreover, $\mathcal{H}$ is an initial segment of $\mathrm{Ult}_0(\mathcal{M}, E)$ as otherwise again we have a cardinal of $L[\dot{E}^{\mathcal{M}} \restriction \rho_\omega^{\mathcal{H}}]$ strictly between $\rho_\omega^{\mathcal{H}}$ and $\mathrm{OR}^{\mathcal{H}}$. So we have alternative (b).

So we may assume $\rho_\omega^{\mathcal{H}} \leq \operatorname{lh} E$ for some E from the $\mathcal{H}$ sequence, and hence $\rho_\omega^{\mathcal{M}} \leq \operatorname{lh} E$ for some E from the $\mathcal{M}$ sequence.

The next section of the the proof will be almost the same as the start of the proof of theorem 8.1. We will compare $\mathcal{H}$ with $\mathcal{M}$ as in in theorem 8.1, with $\rho_\omega^{\mathcal{H}}$ in the place of α_s and ω in the place of k, noticing that the proof of the strong uniqueness theorem gives easily that every ω-maximal iteration tree on $\mathcal{H}$ or $\mathcal{M}$ is simple. Everything will go through almost exactly as before until the point where we used the fact that there was a subset A of α_s which is definable in $\mathcal{H}$ and not in $\mathcal{M}$. Thus we will conclude that $0\bar{T}\theta$, that there is no dropping along $[0,\theta]_{\bar{T}}$, and that $\bar{\mathcal{P}}_\theta \leq \mathcal{Q}_\theta$. It will follow immediately that $\bar{\imath}_{0,\theta}$ is the identity, since the use of any extender with critical point greater than or equal to $\rho_\omega^{\mathcal{H}}$ would cause a drop.

We now continue with the detailed proof. As before, we define three ω-maximal trees by induction on length:

(1) a "psuedo iteration tree" $\bar{\mathcal{T}}$ on the pair $(\mathcal{H}, \mathcal{M})$, with models $\bar{\mathcal{P}}_\alpha$; (2) an iteration tree $\mathcal{T}$ on $\mathcal{M}$ enlarging $\mathcal{T}$, with models $\mathcal{P}_\alpha$, and (3) an iteration tree $\mathcal{U}$ on $\mathcal{M}$ with models $\mathcal{Q}_\alpha$. We also have embeddings

$$\pi_\alpha : \bar{\mathcal{P}}_\alpha \to \mathcal{P}_\alpha$$

such that π_α is a $\overline{\deg}(\alpha)$ embedding. The π_α's have the natural commutativity and agreement properties they had in 8.1.

Set

$$\bar{\mathcal{P}}_0 = \mathcal{H},\ \bar{\mathcal{P}}_{-1} = \mathcal{M},\ \mathcal{P}_0 = \mathcal{Q}_0 = \mathcal{M}$$

and

$$\pi_0 = \pi,\ \pi_{-1} = \text{identity}.$$

The remainder of $\mathcal{T}$, $\bar{\mathcal{T}}$, and $\mathcal{U}$ is defined by induction just as in 8.1: we get $\bar{\mathcal{P}}_{\alpha+1}$ and $\mathcal{Q}_{\alpha+1}$ by "iterating the least disagreement" between $\bar{\mathcal{P}}_\alpha$ and $\mathcal{Q}_\alpha$, as in the comparison process. We get $\pi_{\alpha+1}$ and $\mathcal{P}_{\alpha+1}$ by copying. The role of α_s in the proof of 8.1 is played here by $\rho_\omega^{\mathcal{H}}$; that is, if $\operatorname{crit} \bar{E}_\alpha < \rho_\omega^{\mathcal{H}}$, then $-1\bar{T}(\alpha+1)$.

As before, we get θ such that $\bar{\mathcal{P}}_\theta$ is an initial segment of $\mathcal{Q}_\theta$ or vice-versa.

We say a branch b of $\mathcal{U}$ drops if either $D^{\mathcal{U}} \cap b \neq \emptyset$ or $\deg^{\mathcal{U}}(\alpha) < \omega$ for some $\alpha \in b$. Similarly for branches of $\mathcal{T}$ and $\bar{\mathcal{T}}$. Since we are dealing with ω-maximal trees on fully sound mice, we have that

(a) if $\{\beta \mid \beta\bar{T}\theta\}$ drops, then $\mathcal{Q}_\theta$ is a proper initial segment of $\bar{\mathcal{P}}_\theta$ and $[0,\theta]_{\mathcal{U}}$ doesn't drop and

(b) if $[0,\theta]_{\mathcal{U}}$ drops, then $\bar{\mathcal{P}}_\theta$ is a proper initial segment of $\mathcal{Q}_\theta$ and $\{\beta \mid \beta\bar{T}\theta\}$ doesn't drop.

CLAIM 1. $\{\beta \mid \beta\bar{T}\theta\}$ doesn't drop.

PROOF. By (a) above, if $\{\beta \mid \beta \bar{T} \theta\}$ drops then $\pi_\theta \circ i^{\mathcal{U}}_{0,\theta}$ is a fully elementary embedding from $\mathcal{M}$ to a proper initial segment of $\mathcal{P}_\theta$, which lies on a simple iteration tree based on $\mathcal{M}$. This contradicts the Dodd-Jensen lemma.

CLAIM 2. $0\bar{T}\theta$.

PROOF. Suppose $-1\bar{T}\theta$.

CASE 1. $[0,\theta]_U$ drops. Then $\bar{\imath}_{-1,\theta}$ is a fully elementary embedding from $\mathcal{M}$ to a proper initial segment of Q_θ. This contradicts the Dodd-Jensen lemma.

CASE 2. $[0,\theta]_U$ doesn't drop. If one of $\bar{\mathcal{P}}_\theta$ and Q_θ is a proper initial segment of the other, then we have a contradiction to the Dodd-Jensen lemma. So suppose $\bar{\mathcal{P}}_\theta = Q_\theta$. Then as in Case 3 of the proof of Claim 1 of 8.1, $\bar{\imath}_{-1,\theta} = i^{\mathcal{U}}_{0,\theta}$. This means that the first extender used along $[-1,\theta]_{\mathcal{T}}$ is compatible with the first extender used along $[0,\theta]_{\mathcal{U}}$, which is impossible. □

CLAIM 3. $\bar{\imath}_{0,\theta} =$ identity.

PROOF. Otherwise, since $\rho_\omega^{\bar{\mathcal{P}}_0} \leq \operatorname{crit} \bar{\imath}_{0,\theta}$, $[0,\theta]_{\mathcal{T}}$ drops. This contradicts Claim 1.
□

CLAIM 4. $\bar{\mathcal{P}}_\theta = \mathcal{H}$ is a proper initial segment of Q_θ.

PROOF. If $[0,\theta]_{\mathcal{U}}$ drops, then in fact $\bar{\mathcal{P}}_\theta$ must be a proper initial segment of Q_θ, as $\bar{\mathcal{P}}_\theta$ is ω-sound. If $[0,\theta]_{\mathcal{U}}$ doesn't drop, then $\bar{\mathcal{P}}_\theta$ is an initial segment of Q_θ as otherwise $\pi_\theta \circ i^{\mathcal{U}}_{0,\theta}$ contradicts the Dodd-Jensen lemma. But $\rho_\omega^{\mathcal{H}} < \rho_\omega^{\mathcal{M}} \leq i^{\mathcal{U}}_{0,\theta}(\rho_\omega^{\mathcal{M}}) = \rho_\omega^{Q_\theta}$, so $\bar{\mathcal{P}}_\theta = \mathcal{H} = Q_\theta$ is impossible. □

Our proof now deviates from that of theorem 8.1. In order to show that $\mathcal{U}$ is the desired tree we must verify that either (a) or (b) of the statement of 8.2 holds. Suppose (a) fails, that is, $\mathcal{U}$ is nontrivial. So $E_0^{\mathcal{U}} \neq \emptyset$. Now $\rho_\omega^{\mathcal{H}} \leq \operatorname{lh} E_0^{\mathcal{U}}$ since $\operatorname{crit}(\pi) = \rho_\omega^{\mathcal{H}}$, and $\operatorname{lh} E_0^{\mathcal{U}} \leq \mathrm{OR}^{\mathcal{H}}$, since otherwise $\mathcal{H}$ would be an initial segment of $\mathcal{M}$. But now $\operatorname{lh} E_0^{\mathcal{U}}$ is a cardinal of Q_θ, and $\mathcal{H}$ is a proper initial segment of Q_θ, so that $\operatorname{card}(\mathrm{OR}^{\mathcal{H}}) \leq \rho_\omega^{\mathcal{H}}$ in Q_θ. So we must have $\operatorname{lh} E_0^{\mathcal{U}} = \rho_\omega^{\mathcal{H}}$. Similarly, if $E_1^{\mathcal{U}}$ exists, then $\mathrm{OR}^{\mathcal{H}} < \operatorname{lh} E_1^{\mathcal{U}}$. So in fact $E_1^{\mathcal{U}}$ doesn't exist, that is, $\theta = 1$ and $\mathcal{H}$ is a proper initial segment of $Q_1 = \mathrm{Ult}_k(\mathcal{M}, E_0^{\mathcal{U}})$, where $k = \deg^{\mathcal{U}}(1)$. We can take $k = 0$ because $\mathrm{Ult}_0(\mathcal{M}, E_0^{\mathcal{U}})$ and $\mathrm{Ult}_k(\mathcal{M}, E_0^{\mathcal{U}})$ agree to their common value for $(\rho_\omega^{\mathcal{H}})^+$ and beyond. □

Remark. The hypothesis that $\operatorname{crit}(\pi) = \rho_\omega^{\mathcal{H}}$ is necessary in 8.2. For notice that $\operatorname{crit}(\pi) > \rho_\omega^{\mathcal{H}}$ is impossible since π is fully elementary. (That is, this case is vacuous.) On the other hand, $\operatorname{crit}(\pi) < \rho_\omega^{\mathcal{H}}$ can occur while conclusions (a) and (b) of 8.2 fail: e.g., let $\mathcal{M} = \mathrm{Ult}_\omega(\mathcal{H}, E)$ where E is on the $\mathcal{H}$ sequence and $\operatorname{crit}(E) < \rho_\omega^{\mathcal{H}}$, and let π be the canonical embedding.

One can also derive a version of 8.2 with $\rho_{n+1}^{\mathcal{H}}$ replacing $\rho_\omega^{\mathcal{H}}$. Namely, suppose $\mathcal{H}$ and $\mathcal{M}$ are 1-small, $n+1$ sound mice, and $\pi : \mathcal{H} \to \mathcal{M}$ is $r\Sigma_{n+1}$ elementary with $\operatorname{crit}(\pi) \geq \rho_{n+1}^{\mathcal{H}}$. Then either

(a) $\mathcal{H}$ is a proper initial segment of $\mathcal{M}$, or
(b) $\rho^{\mathcal{H}}_{n+1} = \operatorname{lh} E$ for some E from the $\mathcal{M}$ sequence, and $\mathcal{H}$ is a proper initial segment of $\mathrm{Ult}_0(\mathcal{M}, E)$.

The example following the statement of 8.2 shows alternative (b) is necessary.

The proof of this version is almost the same as that of 8.2. We use n-maximal trees in the comparison and modify the uses of Dodd-Jensen slightly to accommodate this change. Note that in this case we don't know that if e.g. $[0, \theta]_{\mathcal{U}}$ drops then $\bar{\mathcal{P}}_\theta$ is a *proper* initial segment of Q_θ. Also notice that we can assume that there is an extender E from $\mathcal{M}$ sequence with $\operatorname{lh}(\mathcal{M}) \geq \rho^{\mathcal{M}}$, since the result is trivial otherwise.

Notice that alternative (b) of 8.2 (or its "$n+1$ version") cannot arise when $\rho^{\mathcal{H}}_\omega$ (respectively $\rho^{\mathcal{H}}_{n+1}$) is a cardinal of $\mathcal{M}$, simply because $\operatorname{lh} E$ is never a cardinal of $\mathcal{M}$ when E is on the $\mathcal{M}$ sequence and $\operatorname{lh} E < \mathrm{OR}^{\mathcal{M}}$.

As a sample application of the $n+1$-version: let $\mathcal{M}$ be a 1-small, 1-sound mouse, and let $\alpha < \rho^{\mathcal{M}}_1$, α a cardinal of $\mathcal{M}$. Let $p = p_1(\mathcal{M})$, and $\mathcal{H} = \mathcal{H}^{\mathcal{M}}_1(\alpha \cup \{p\})$. Let $\pi : \mathcal{H} \to \mathcal{M}$ be the inverse of the collapse. Clearly $\alpha = \rho^{\mathcal{H}}_1 \leq \operatorname{crit}(\pi)$, and π is $r\Sigma_1$ elementary. Suppose α is large enough that the solidity witnesses for p are all of the form $\tau^{\mathcal{M}}[\bar{\beta}, p]$ for some $\bar{\beta} \in \alpha^{<\omega}$ and $\tau \in \mathrm{Sk}_1$. This guarantees that $\pi^{-1}(p)$ is the first standard parameter of $\mathcal{H}$, and that $\mathcal{H}$ is 1-sound. We can then conclude that $\mathcal{H}$ is a proper initial segment of $\mathcal{M}$.

We don't know whether the assumption that $\mathcal{H}$ is $n+1$ sound can be reduced to n soundness. If this can be done, then in the application just mentioned we needn't assume $p = p_1(\mathcal{M})$ or make the largeness assumption about α.

§9. Uniqueness of the Next Extender

In §11 we shall construct an extender sequence $\vec{E}$ such that $L[\vec{E}] \models$ "there is a Woodin cardinal, and every level $\mathcal{J}_\alpha^{\vec{E}}$ of $L[\vec{E}]$ is a 1-small coremouse". The sequence $\vec{E}$ will be defined by recursion. The recursion is substantially more subtle than it is for sequences of measures, but the basic idea is still to define $\vec{E}_\gamma$ by recursion on γ, by making $\vec{E}_\gamma$ be the least extender which can be added to the sequence $\vec{E} \upharpoonright \gamma$ so that the extender sequence remains good. Part of the strategy will be to pick $\vec{E}_\gamma$ without regard to the initial segment condition and then prove that in fact it does satisfy the initial segment condition as well. We would like to show that there is always only one possible choice of $\vec{E}_\gamma$ for each γ, so that if ρ is the natural length of $\vec{E}_\gamma \upharpoonright \rho$ and G of length γ' is its trivial extension then G, being a legal choice for $E_{\gamma'}$, must in fact be $E_{\gamma'}$. Of course this ignores the second alternative in the initial segment condition, but more important we are unable to prove this uniqueness: so far as we know there could be one choice of types I or III and a second of type II. In this section we will prove uniqueness for types I or III, and in section 11 this will be used for the case when ρ is a cardinal in $L[\vec{E}]$. In section 10 we will prove a related result which will apply in the cases when ρ is not a cardinal in $L[\vec{E}]$.

The standard method for showing uniqueness of the next extender on the sequence involves *Doddages* and comparison of a Doddage with itself. The method originates in Mitchell's [M74R], see also [D]. We need only a simple sort of Doddage, dubbed by Jensen a bicephalus. A bicephalus is like an active premouse, except that it has two predicates corresponding to two candidates for a last extender. By comparing bicephali with themselves we show that in sufficiently iterable bicephali, these candidates are not distinct.

Unfortunately, when we want to form an ultrapower of a bicephalus whose last extenders differ in type, we have a problem. We may want to squash for the sake of one extender, but if we do so it is not clear how to carry along the other. This is the reason we will also need the alternative technique from section 10.

The first problem in dealing with bicephali will be to verify that when we form the ultrapower of a bicephalus both of whose last extenders are of type III, the squashing procedures in the two cases are consistent with one another. We shall verify this now, in Lemma 9.1.

If M is an active ppm then $\nu^{\mathcal{M}}$ is just the the natural length of the extender coded by $\dot{F}^{\mathcal{M}}$, that is if $\mathcal{M}$ is of type II or III then ν^{M} is the strict sup of its generators, while if $\mathcal{M}$ is type I, then $\nu^{\mathcal{M}} = (\kappa^+)^{\mathcal{M}}$.)

Lemma 9.1. *Let $\mathcal{M}$ be a type III ppm, and G an extender over M with* $\operatorname{crit} G = \kappa < \nu^{\mathcal{M}}$. *Let $\mathcal{P}$ be the ultrapower of $\mathcal{M}$ via G, where functions in $|\mathcal{M}|$ are used, and let $i : \mathcal{M} \to \mathcal{P}$ be the canonical embedding. Assume $\mathcal{P}$ is wellfounded. Let $\nu^* = sup\ i''\nu^{\mathcal{M}}$. Then*

(a) $\nu^* =$ *sup of generators of* $\dot{F}^{\mathcal{P}}$.

(b) *Let* $\gamma = (\nu^*)^{+^{\mathcal{P}}}$, *or* $\gamma = OR^{\mathcal{P}}$ *if* $\mathcal{P} \models (\nu^*)^+$ *doesn't exist. Let*

$$Q = (J_\gamma^{\dot{E}^{\mathcal{P}}}, \in, \dot{E}^{\mathcal{P}} \restriction \gamma, \dot{F}^{\mathcal{P}} \restriction \gamma).$$

Then Q *is a type* III *ppm. and* $Q^{sq} = \mathrm{Ult}_0(\mathcal{M}^{sq}, G)$.

Remarks. (1) $\mathcal{P}$ is defined more carefully at the beginning of §4. It is to be constructed without squashing.

(2) According to (b), the structure Q is equal to the structure $\mathrm{Ult}_0(\mathcal{M}, G)$.

(3) If $\nu^* = i(\nu^{\mathcal{M}})$, then $\gamma = \mathrm{OR}^{\mathcal{P}}$, so that $\mathcal{P} = Q$.

(4) If $\nu^* < i(\nu^{\mathcal{M}})$, then $\gamma < \mathrm{OR}^{\mathcal{P}}$ since $i(\nu^{\mathcal{M}})$ is a cardinal of $\mathcal{P}$. Part (a) of 9.1 then tells us that $\mathcal{P}$ is not a ppm, as it violates the bounded generators clause of "good at $\mathrm{OR}^{\mathcal{P}}$". (It also violates the initial segment clause.) Roughly speaking, its last extender $\dot{F}^{\mathcal{P}}$ was added too late; it should have been added at γ. Replacing $\mathcal{P}$ by Q, which is the net effect of squashing, amounts to adding $\dot{F}^{\mathcal{P}}$ at γ.

Proof of 9.1. (a) Notice that for $\xi < \nu^{\mathcal{M}}$, ξ is a generator of $\dot{F}^{\mathcal{M}}$ iff $i(\xi)$ is a generator of $\dot{F}^{\mathcal{P}}$. The reason is that $\dot{F}^{\mathcal{M}} \restriction (\xi+1) \in |\mathcal{M}|$, and the fact that ξ is a generator is a Σ_0 fact about $\dot{F}^{\mathcal{M}} \restriction (\xi+1)$. Thus ν^* is a sup of generators of $\dot{F}^{\mathcal{P}}$, and we need only show that no $\eta \geq \nu^*$ is a generator of $\dot{F}^{\mathcal{P}}$.

So let $\nu^* \leq \eta < \mathrm{OR}^{\mathcal{P}}$. We want to find $a \subseteq \nu^*$, a finite, and $h \in |\mathcal{P}|$ such that $\eta = [a, h]^{\mathcal{P}}_{\dot{F}^{\mathcal{P}}}$.

Let $\eta = [b, f]^{\mathcal{M}}_G$, where b is a size n set of coordinates of G and $f \in |\mathcal{M}|$ and $\mathrm{dom}\, f = [\kappa]^n$. It will be enough to find maps $\bar{u} \mapsto a_{\bar{u}}$ and $\bar{u} \mapsto h_{\bar{u}}$, both in $|\mathcal{M}|$, such that

(i) for G_b a.e. $\bar{u}$, $f(\bar{u}) = [a_{\bar{u}}, h_{\bar{u}}]^{\mathcal{M}}_{\dot{F}^{\mathcal{M}}}$

and

(ii) $\bigcup\{\, a_{\bar{u}} : \bar{u} \in [\kappa]^n \,\}$ is bounded in $\nu^{\mathcal{M}}$.

(We can then take $a = [b, \lambda\bar{u}\cdot a_{\bar{u}}]^{\mathcal{M}}_G$ and $h = [b, \lambda\bar{u}\cdot h_{\bar{u}}]^{\mathcal{M}}_G$. Because $\dot{F}^{\mathcal{M}}$ is weakly amenable over $\mathcal{M}$ we have enough of Los' theorem to show this works. We omit the details.)

Now as $f \in |\mathcal{M}|$, the coherence condition on $\dot{F}^{\mathcal{M}}$ implies that $f \in \mathrm{Ult}(\mathcal{M}, \dot{F}^{\mathcal{M}})$. Let

$$f = [c, g]^{\mathcal{M}}_{\dot{F}^{\mathcal{M}}}$$

where $c \subseteq \nu^{\mathcal{M}}$ is finite and $g \in |\mathcal{M}|$. For any $\bar{u} = \{u_1 \cdots u_n\} \in [\kappa]^n$, set

$$a_{\bar{u}} = c \cup \{u_1 \cdots u_n\}.$$

So the map $\bar{u} \mapsto a_{\bar{u}}$ is in $\mathcal{M}$, and condition (ii) above holds.

Let $a_{\bar{u}} = \{\alpha_1 \cdots \alpha_I\}$ in increasing order. Let $c = \{\alpha_{k_0} \cdots \alpha_{k_e}\}$ and $\bar{u} = \{\alpha_{m_0} \cdots \alpha_{m_i}\}$ in increasing order. If $\bar{v} = \{v_1 \cdots v_I\}$ is any sequence, given in increasing order, then we will write $\bar{v}^0_{\bar{u}} = \{v_{k_0} \cdots v_{k_e}\}$ and $\bar{v}^1_{\bar{u}} = \{v_{m_0} \cdots v_{m_i}\}$. Note that the map $(\bar{v}, \bar{u}) \mapsto (\bar{v}^0_{\bar{u}}, \bar{v}^1_{\bar{u}})$ is in M. Now set, for $\bar{u} \in [\kappa]^n$ and $\bar{v}$ in the space of $\dot{F}^{\mathcal{M}}_{a_e}$

$$h_{\bar{u}}(\bar{v}) = g(\bar{v}^0_{\bar{u}})(\bar{v}^1_{\bar{u}}) .$$

(Take $h_{\bar{u}}(\bar{v}) = 0$ if this doesn't make sense; it makes sense $\dot{F}^{\mathcal{M}}_{a_e}$ a.e.) Clearly the map $\bar{u} \mapsto h\bar{u}$ is in $|\mathcal{M}|$. We need only verify condition (i).

Let $j : \mathcal{M} \to \mathrm{Ult}(\mathcal{M}, \dot{F}^{\mathcal{M}})$ be the canonical embedding. Then

$$\begin{aligned}[a_{\bar{u}}, h_{\bar{u}}]^{\mathcal{M}}_{\dot{F}^{\mathcal{M}}} &= j(h_{\bar{u}})(a_{\bar{u}}) \\ &= j(g)((a_{\bar{u}})^0_{\bar{u}})((a_{\bar{u}})^1_{\bar{u}}) \\ &= j(g)(c)(\bar{u}) = f(\bar{u}) ,\end{aligned}$$

as desired.

(b) Clearly, $\mathrm{Ult}_0(\mathcal{M}^{sq}, G) = (J^{\dot{E}^{\mathcal{P}}}_{\nu^*}, \in, \dot{E}^{\mathcal{P}} \restriction \nu^*, \dot{F}^{\mathcal{P}} \restriction \nu^*)$. This structure is $\mathcal{R}^{sq}$ for some type III ppm $\mathcal{R}$, by results of §3. We decode $\mathcal{R}_{\mathcal{P}}$ from $\mathcal{R}^{sq}$ by taking $\mathrm{Ult}_0(\mathcal{R}^{sq}, \dot{F}^{\mathcal{P}} \restriction \nu^*)$ and cutting off at its $(\nu^*)^+$. By (a), this is the same as taking $\mathrm{Ult}_0(R^{sq}, \dot{F}^{\mathcal{P}})$ and cutting off. As $\dot{F}^{\mathcal{P}}$ coheres with the $\mathcal{P}$ sequence, this gives us Q. □

DEFINITION 9.1.1. A *bicephalus* is a structure

$$\mathfrak{B} = (J^{\vec{E}}_{\alpha}, \in, \vec{E} \restriction \alpha, F, G)$$

such that $\mathfrak{B}_0 = (J^{\vec{E}}_{\alpha}, \in, \vec{E} \restriction \alpha, F)$ and $\mathfrak{B}_1 = (J^{\vec{E}}_{\alpha}, \in, \vec{E} \restriction \alpha, G)$ are both premice, and either

(a) both $\mathfrak{B}_0$ and $\mathfrak{B}_1$ are of type II

or

(b) neither $\mathfrak{B}_0$ nor $\mathfrak{B}_1$ is of type II.

Remark. As the reader may have noticed, the distinction between types I and II is not very important elsewhere - here it is.

If both $\mathfrak{B}_0$ and $\mathfrak{B}_1$ are of type II we say $\mathcal{L}$ has type II. Otherwise $\mathfrak{B}$ has type III.

We let $\dot{F}^{\mathfrak{B}}_0$ and $\dot{F}^{\mathfrak{B}}_1$ be the two last extenders of $\mathfrak{B}$.

Certain notions appropriate for premice - e.g. $\mathcal{J}^{\mathfrak{B}}_{\gamma}$, agreement below γ- extend to bicephali in an obvious way.

Suppose $\mathfrak{B}$ and $\mathcal{A}$ are bicephali, and G is an extender from the $\mathcal{A}$-sequence such that crit $G = \kappa$. Suppose $P(\kappa) \cap |\mathfrak{B}| = P(\kappa) \cap |\mathcal{A}|$. Suppose also that if $\mathfrak{B}$ is type III, then $\kappa < \nu^{\mathfrak{B}_0}$. (Notice that $\nu^{\mathfrak{B}_0} = \nu^{\mathfrak{B}_1}$ = largest cardinal of $\mathfrak{B}$, in the case $\mathfrak{B}$ is of type III.) Suppose $\mathrm{Ult}_0(\mathfrak{B}_0, G)$ and $\mathrm{Ult}_0(\mathfrak{B}_1, G)$ are wellfounded, and hence premice of the same type as $\mathfrak{B}_0$ and $\mathfrak{B}_1$. We claim there is a unique bicephalus $\mathcal{C}$ such that $\mathcal{C}_0 = \mathrm{Ult}_0(\mathfrak{B}_0, G)$ and $\mathcal{C}_1 = \mathrm{Ult}_0(\mathcal{B}_1, G)$. If $\mathfrak{B}$ is of type II this is obvious, so suppose $\mathfrak{B}$ is of type III.

Suppose one of $\mathfrak{B}_0$ and $\mathfrak{B}_1$ in of type I. If both are type I, there is no problem. Suppose e.g. $\mathfrak{B}_0$ is type I while $\mathfrak{B}_1$ is type III. Then $\nu^{\mathfrak{B}_1} = (\kappa^+)^{\mathfrak{B}}$, where κ is the critical point of the last extender of $\mathfrak{B}_0$, i.e. of $\dot{F}_0^{\mathfrak{B}}$. (For $\nu^{\mathfrak{B}_1}$ is the largest cardinal of $\mathfrak{B}_1$ in the type III case, and $(\kappa^+)^{\mathfrak{B}_0}$ is the largest cardinal of $\mathfrak{B}_0$ in the type I case.) But then if i is the canonical embedding from the full ultrapower of $\mathfrak{B}_1$ by G, using all functions in $|\mathfrak{B}_1|$, then i is continuous at $\nu^{\mathfrak{B}_1}$. So by Lemma 9.1, the squashed and unsquashed ultrapowers of $\mathfrak{B}_1$ coincide. This gives us the desired $\mathfrak{C}_1$ at once.

Now suppose both $\mathfrak{B}_0$ and $\mathfrak{B}_1$ are of type III. Recall that if $\mathcal{M}$ is a type III premouse, then $\mathrm{Ult}_0(\mathcal{M}, G)$ is the unique Q such that $Q^{sq} = \mathrm{Ult}_0(\mathcal{M}^{sq}, G)$. It will be enough to show that $\mathrm{OR} \cap \mathrm{Ult}_0(\mathcal{B}_0, G) = \mathrm{OR} \cap \mathrm{Ult}_0(\mathcal{B}_1, G)$, and that $\mathrm{Ult}_0(\mathfrak{B}_0, G)$ agrees with $\mathrm{Ult}_0(\mathfrak{B}_1, G)$ below $\mathrm{OR} \cap \mathrm{Ult}_0(\mathcal{B}_0, G)$. But now let $\mathcal{D}$ be the full ultrapower via G of $\mathcal{B}$ formed using all functions in $|\mathcal{B}|$, and $i : \mathcal{B} \to \mathcal{D}$ the canonical embedding. By Lemma 9.1 we see $\mathrm{OR} \cap \mathrm{Ult}_0(\mathcal{B}_0, G) = (\sup i''\nu^{\mathcal{B}_0})^{+\mathcal{D}} = \mathrm{OR} \cap \mathrm{Ult}_0(\mathcal{B}_1, G)$, and that the necessary agreement holds.

So we may define

DEFINITION 9.1.2. In the situation described above, $\mathrm{Ult}_0(\mathfrak{B}, G)$ is the unique bicephalus $\mathfrak{C}$ such that $\mathfrak{C}_0 = \mathrm{Ult}_0(\mathfrak{B}_0, G)$ and $\mathfrak{C}_1 = \mathrm{Ult}_0(\mathfrak{B}_1, G)$.

Notice that if $\mathfrak{B}$ is type II, we have a canonical $i : \mathfrak{B} \to \mathrm{Ult}_0(\mathfrak{B}, G)$ which is $r\Sigma_1$ elementary (in the obvious sense.) If $\mathfrak{B}$ is type III we get an embedding $i : \mathfrak{B}^{sq} \to \mathrm{Ult}_0(\mathfrak{B}, G)^{sq}$ - which is $q\Sigma_1$ elementary. We get an embedding $i : \mathfrak{B} \to \mathrm{Ult}_0(\mathfrak{B}, G)$ in the case $\mathrm{Ult}_0(\mathfrak{B}, G)$ happens to be the full ultrapower of $\mathfrak{B}$ by G, using all functions in $|\mathfrak{B}|$. This happens when the canonical embedding of $\mathfrak{B}$ into the full ultrapower, call it i, is continuous at $\nu^{\mathfrak{B}_0}$. That is, this happens when $\mathfrak{B} \models \mathrm{cof}\,(\nu^{\mathfrak{B}_0}) \neq \kappa$, where $\kappa = \mathrm{crit}\ G$. Notice in this regard that if $\mathfrak{B} \models \mathrm{cof}\,(\nu^{\mathfrak{B}_0}) = \kappa$, and $\mathfrak{C} = \mathrm{Ult}_0(\mathfrak{B}, G)$, then $\mathfrak{C} \models \mathrm{cof}\,(\nu^{\mathfrak{C}_0}) = \kappa$, since $\nu^{\mathfrak{C}_0} = \sup i''\, \nu^{\mathfrak{B}_0}$. This implies that along any branch of an iteration tree on $\mathfrak{B}$, the natural embeddings map $\mathfrak{B}$ into $\mathrm{Ult}_0(\mathfrak{B}, G)$ in all but at most one instance. This is because we can hit a given κ at most once along any branch.

The notion of a 0-maximal iteration tree generalizes in an obvious way to trees on bicephali, so we shall just mention a few points. Let $\mathfrak{B}$ be a bicephalus; a 0-maximal iteration tree on $\mathfrak{B}$ is a system

$$\mathcal{T} = \langle T, \deg, D, \langle E_\alpha, \mathfrak{B}^*_{\alpha+1} \mid \alpha + 1 < \theta \rangle \rangle$$

together with associated models $\mathfrak{B}_\alpha$ and embeddings $i_{\alpha\beta} : \mathfrak{B}_\alpha \to \mathfrak{B}_\beta$ defined whenever $\alpha T \beta$ and $[\alpha, \beta]_T \cap D = \emptyset$. We have $\mathfrak{B}_0 = \mathfrak{B}$. One can build $\mathcal{T}$ freely at successor steps except for the following restrictions:

Let T-pred$(\alpha + 1) = \beta$, and crit $E_\alpha = \kappa$. Then we must have $\mathrm{lh}\, E_\alpha > \mathrm{lh}\, E_\eta$, all $\eta < \alpha$, and

$$\beta = \text{least } \xi \leq \alpha \text{ such that } \kappa < \text{sup of generators of } E_\xi .$$

Let

$$\gamma = \text{largest } \eta \text{ such that } \mathcal{J}_\eta^{\mathfrak{B}_\beta} \text{ exists and}$$
$$P(\kappa) \cap |\mathcal{J}_\eta^{\mathfrak{B}_\beta}| \subseteq |\mathfrak{B}_\alpha| ;$$

then $\alpha + 1 \in D \Leftrightarrow \mathcal{J}_\gamma^{\mathfrak{B}_\beta}$ is a proper initial segment of $\mathfrak{B}_\beta$, and

$$\deg(\alpha+1) = \begin{cases} 0 & \text{if } [0, \alpha+1]_T \cap D = \emptyset , \\ \text{largest } k \text{ s.t. } \kappa < \rho_k^{\mathcal{N}}, \text{ for } \mathcal{N} = \mathcal{J}_\gamma^{\mathfrak{B}_\beta} & \text{, otherwise} \end{cases}$$

and

$$\mathfrak{B}^*_{\alpha+1} = \mathcal{J}_\gamma^{\mathfrak{B}_\beta} .$$

Finally, if $n = \deg(\alpha + 1)$, then

$$\mathfrak{B}_{\alpha+1} = \mathrm{Ult}_n(\mathfrak{B}^*_{\alpha+1}, E_\alpha)$$

and if $\alpha + 1 \notin D$ we have a canonical embedding $i_{\beta,\alpha+1} : \mathfrak{B}_\beta \to \mathfrak{B}_{\alpha+1}$.

This last statement is not true in the case when $\mathfrak{B}_\beta$ is of type III and $\nu^{\mathfrak{B}_\beta}$ has cofinality κ in $\mathfrak{B}_\beta$. In this case we let $i_{\beta,\alpha+1}$ be the canonical embedding of $\mathfrak{B}_\beta^{sq}$ into $\mathfrak{B}^{sq}_{\alpha+1}$.]

We also have an embedding $i^*_{\alpha+1} \to \mathfrak{B}_{\alpha+1}$, again with a possible exception in the type III case, when we may only have $i^*_{\alpha+1} : (\mathfrak{B}^*_{\alpha+1})^{sq} \to (\mathfrak{B}_{\alpha+1})^{sq}$.

If $\lambda < \theta$ is a limit, then $D \cap [0, \lambda]_T$ must be finite. Moreover the special case referred to above will only occur finitely often, so that $\mathrm{dom}\, i^*_{\alpha+1} = |\mathfrak{B}^*_{\alpha+1}|$ for all but finitely many $\alpha + 1 \in [0, \lambda)_T$. Thus the direct limit of the models $\mathfrak{B}_\beta$ under the maps $i_{\beta,\gamma}$ for β, γ in $[\beta_0, \lambda)$ for some $\beta_0 \in [0, \lambda)$ exists; and we require that $\mathfrak{B}_\lambda$ be this direct limit.

Remarks.

1. If $[0, \alpha + 1]_T \cap D \neq \emptyset$, then $\mathfrak{B}_{\alpha+1}$ is a premouse rather than a bicephalus. Moreover, one can see by an easy induction that $\mathfrak{B}^*_{\alpha+1}$ is $\deg(\alpha + 1)$ sound, whenever $[0, \alpha + 1]_T \cap D \neq \emptyset$. Also, if $\gamma + 1 T \alpha + 1$ and$D \cap (\gamma + 1, \alpha + 1]_T = \emptyset$, then $\deg(\gamma + 1) \geq \deg(\alpha + 1)$.

2. By coherence and the fact that $\mathrm{lh}\, E_\alpha$ increases with α, we get the counterpart of Lemma 5.1; $\mathfrak{B}_\beta$ agrees with $\mathfrak{B}_\alpha$ below $\mathrm{lh}\, E_\alpha$, for all $\beta \geq \alpha$, and $\mathrm{lh}\, E_\alpha$ is a cardinal of $\mathfrak{B}_\beta$ for all $\beta > \alpha$.

The notion of a *simple* iteration tree generalizes in an obvious way to trees on bicephali. We can then define *k-iterability* for bicephali just as we did for premice.

The notion of 1-*smallness* generalizes in the obvious way: a bicephalus $\mathfrak{B}$ is 1-small iff both $\mathfrak{B}_0$ and $\mathfrak{B}_1$ are 1-small. The *uniqueness theorem* 6.1 generalizes in an obvious way. (We have no analog of 6.2, strong uniqueness. In general, we don't care about the Levy hierarchy over bicephali.)

The main theorem about bicephali is that there aren't any interesting ones.

Theorem 9.2. *Let $\mathfrak{B}$ be a 1-small, 1-iterable bicephalus. Then $\dot{F}_0^{\mathfrak{B}} = \dot{F}_1^{\mathfrak{B}}$.*

PROOF SKETCH. We compare $\mathfrak{B}$ with itself in such a way that the comparison process can only terminate if $\dot{F}_0^{\mathfrak{B}} = \dot{F}_1^{\mathfrak{B}}$. But iterability implies that the process terminates, so $\dot{F}_0^{\mathfrak{B}} = \dot{F}_1^{\mathfrak{B}}$.

Let $\mathcal{T}$ and $\mathcal{U}$ be the 0-maximal iteration trees on $\mathfrak{B}$, with models $\mathfrak{C}_\alpha$ and $\mathcal{D}_\alpha$, built by the method of "iterating the least disagreement". Notice that if $\dot{F}_0^{\mathfrak{B}} \neq \dot{F}_1^{\mathfrak{B}}$ then it is guaranteed that there will be such a disagreement, since even if the bicephalus $\mathfrak{C}_\alpha = (J_\gamma^{\vec{E}}, \in, \vec{E}, F, G)$ is matched to $\mathcal{D}_\alpha$ except for the final extenders F and G then each of F and G would have to agree with whatever is on the $\mathcal{D}_\alpha$ sequence at γ (or possibly to both of the final extenders of $\mathcal{D}_\alpha$), and if F and G are different then they cannot both agree with the same extender.

At limit steps λ, we stop the construction unless $\mathcal{T} \restriction \lambda$ and $\mathcal{U} \restriction \lambda$ are simple. In the latter case, we let $[0, \lambda)_{\mathcal{T}}$ and $[0, \lambda)_{\mathcal{U}}$ be the unique cofinal wellfounded branches of their respective trees, and continue.

Suppose the construction never stops because we reach a λ such that one of $\mathcal{T} \restriction \lambda$ and $\mathcal{U} \restriction \lambda$ is not simple. Then the proof of 7.1 shows that we must reach a θ such that $\mathfrak{C}_\theta$ is an initial segment of $\mathcal{D}_\theta$, or vice-versa. (So the construction stops at θ.) Say $\mathfrak{C}_\theta$ is an initial segment of $\mathcal{D}_\theta$. By the proof of 7.1, there's no dropping on $[0, \theta]_{\mathcal{T}}$. [Otherwise $\mathfrak{C}_\theta$ is unsound, so $\mathfrak{C}_\theta = \mathcal{D}_\theta$. So $\mathfrak{C}_\omega(\mathfrak{C}_\theta)$, the core of $\mathfrak{C}_\theta$, is $\mathfrak{C}^*_{\alpha+1}$ for some $\alpha + 1 \in [0, \theta)_{\mathcal{U}}$. This is too much agreement at an earlier stage.] Thus $\mathfrak{C}_\theta$ is a bicephalus. If $\dot{F}_0^{\mathfrak{C}_\theta} \neq \dot{F}_1^{\mathfrak{C}_\theta}$, then $\mathfrak{C}_\theta$ cannot be on initial segment of $\mathcal{D}_\theta$; at worst, one of $\dot{F}_0^{\mathfrak{C}_\theta}$ and $\dot{F}_1^{\mathfrak{C}_\theta}$ will participate in a disagreement with $\mathcal{D}_\theta$. So $\dot{F}_0^{\mathfrak{C}_\theta} = \dot{F}_1^{\mathfrak{C}_\theta}$, so $\dot{F}_0^{\mathfrak{B}} = \dot{F}_1^{\mathfrak{B}}$.

Suppose we reach a λ s.t. e.g. $\mathcal{T} \restriction \lambda$ is not simple. Let b and c be distinct cofinal wellfounded branches of $\mathcal{T} \restriction \lambda$. Suppose e.g. $\mathrm{OR}^{\mathfrak{C}_b} \leq \mathrm{OR}^{\mathfrak{C}_c}$. Let $\delta = \delta(\mathcal{T} \restriction \lambda) = \sup\ \{\mathrm{lh}\, E_\alpha^{\mathcal{T}} \mid \alpha \in \lambda\}$. The proof of Claim 1 in the proof of 6.2 shows $\mathrm{lh}\, F < \delta$ for all extenders F from the $\mathfrak{C}_b$ sequence. Clearly, then, $\delta = \sup\ \{\mathrm{lh}\, F : F \text{ from the } \mathfrak{C}_b \text{ sequence}\}$. As $\mathfrak{C}_0 = \mathfrak{B}$ has a maximum length realized by extenders on its sequence, $D^{\mathcal{T}} \cap b \neq \varnothing$. Thus $\mathfrak{C}_b$ is unsound. On the other hand, the proof of Claim 2 in/a the proof of 6.2 shows $\mathfrak{C}_b$ is an initial segment of $\mathfrak{C}_c$. Thus $\mathfrak{C}_b = \mathfrak{C}_c$, and $D^{\mathcal{T}} \cap c \neq \varnothing$. But then the proof of Claim 4 in the proof of 6.2 yields a contradiction.

Thus we can never reach a λ such that one of $\mathcal{T} \upharpoonright \lambda$ and $\mathcal{U} \upharpoonright \lambda$ is not simple. This completes the proof. □

We prove a result, theorem 10.1, which implies that certain structures arising in the construction done in §11 satisfy the initial segment condition on premice. As was pointed out in the last section, this will be used when bicephali cannot be used because one of the extenders being compared is of type II and one is not.

Definition 10.0.1. A psuedo-premouse is a structure $\mathcal{M} = (J_\alpha^{\vec{E}}, \in, \vec{E}, \tilde{F})$ such that

(1) $(J_\alpha^{\vec{E}}, \in, \vec{E})$ is a passive premouse,
(2) F satisfies conditions 1 through 4 in the definition of "good at α" (i.e. everything but the initial segment condition), and
(3) There is a $\delta < \alpha$ s.t. (i) $\mathcal{M} \models \delta$ is the largest cardinal and (ii) for some γ s.t. $\delta < \gamma < \alpha$, $\gamma \in \text{dom}\, \vec{E}$ and $E_\gamma =$ trivial completion of $F \restriction \delta$.

Any psuedo-premouse $\mathcal{M}$ is weakly amenable with respect to its predicate $\dot{F}^{\mathcal{M}}$ for the last extender. Consequently, if E is an extender from the sequence of some psuedo-premouse $\mathcal{N}$, then we can define $\text{Ult}_0(\mathcal{M}, E)$ in the natural way, as for premice. If $\text{Ult}_0(\mathcal{M}, E)$ is wellfounded, we identify it with its transitive collapse. Los' theorem holds for $r\Sigma_0$ formulae and so if $i : \mathcal{M} \to \text{Ult}_0(\mathcal{M}, E)$ is the canonical embedding, i is $r\Sigma_1$ elementary. The calculations of §2 show that, if transitive, $\text{Ult}_0(\mathcal{M}, E)$ is a psuedo-premouse. (If δ, γ witness 3 for $\mathcal{M}$, then $i(\delta)$, $i(\gamma)$ witness 3 for $\text{Ult}_0(\mathcal{M}, E)$.) We can thus construct 0-maximal iteration trees on a psuedo-premouse $\mathcal{M}$. We define the notions of simplicity and iterability for psuedo-premice just as for premice. (We only consider 0-maximal trees.) The notion of 1-smallness also generalizes in an obvious way.

Theorem 10.1. *Let $\mathcal{M}$ be an iterable, 1-small psuedo-premouse. Then $\mathcal{M}$ is a premouse.*

Proof. We must show that the initial segment condition holds. Let $\mathcal{M} = (J_\alpha^{\vec{E}}, \in, \vec{E}, \tilde{F})$, and suppose toward a contradiction that the initial segment condition fails for $F \restriction \rho$. Thus ρ is the natural length of $F \restriction \rho$ and if G is the trivial completion of $F \restriction \rho$ then G is not on the $\dot{\vec{E}}^{\mathcal{M}}$ sequence, and if $\rho \in \text{dom}\, \dot{\vec{E}}^{\mathcal{M}}$ then G is not on the $\dot{E}^{\text{Ult}(\mathcal{M}, \dot{E}_\rho^{\mathcal{M}})}$ sequence.

Notice that if ρ is a successor ordinal then $\rho - 1$ is a generator of F, and if ρ is a limit ordinal then either ρ is a limit of generators of F or else ρ is equal to κ^+ of $\mathcal{M}$ where $\kappa = \text{crit}(F)$. Also, ρ is smaller than natural length of F and as $\mathcal{M}$ is a psuedo-premouse ρ is larger than any cardinal of $\mathcal{M}$.

We obtain a contradiction by comparing $\mathcal{M}$ with $\text{Ult}_0(\mathcal{M}, G)$. That is, we define 0-maximal iteration trees $\mathcal{T}$ and $\mathcal{U}$ on $\mathcal{M}$ with models $\mathcal{P}_\alpha$ and $\mathcal{Q}_\alpha$ respectively as follows:

$$\mathcal{P}_0 = \mathcal{Q}_0 = \mathcal{M}$$
$$\mathcal{P}_1 = \mathrm{Ult}_0(\mathcal{M}, G)$$
$$\mathcal{Q}_1 = \begin{cases} \mathrm{Ult}_0(\mathcal{M}), \dot{E}^{\mathcal{M}}_{\rho} & \text{if } \rho \in \mathrm{dom}\ \dot{E}^{\mathcal{M}} \\ \mathcal{M} & \text{if } \rho \notin \mathrm{dom}\ \dot{E}^{\mathcal{M}} \end{cases}$$

Thus $E^{\mathcal{T}}_0$ is always equal to G, and $E^{\mathcal{U}}_0$ either does not exist or is equal to $\dot{E}^{\mathcal{M}}_{\rho}$.

The remainder of the trees $\mathcal{T}$ and $\mathcal{U}$ is determined by the comparison process. At successor steps we pick an extender, or two extenders, representing the least disagreement, and apply these to possibly earlier models in their respective trees so as not to move generators along branches of $\mathcal{T}$ and $\mathcal{U}$. At limit steps we use the unique cofinal wellfounded branches of $\mathcal{T}$ and $\mathcal{U}$ given by the iterability and 1-smallness of $\mathcal{M}$.

First we will verify that the iteration stops, that is, that there is an ordinal θ such that the θth model $\mathcal{P}_\theta$ of $\mathcal{T}$ is an initial segment of the θth model $\mathcal{Q}_\theta$ of $\mathcal{U}$ or vice-versa. There is a slight wrinkle here because the proof that the comparison process terminates uses the initial segment condition on premice, and we don't yet know that this holds for the final extender F of $\mathcal{M}$.

Suppose that the iteration never stops. As in the proof of the comparison lemma (7.1) we have ordinals $1 \leq \alpha < \beta$ such that $E^{\mathcal{T}}_\alpha$ is the trivial completion of $E^{\mathcal{U}}_\beta \restriction \rho^{\mathcal{T}}_\alpha$ (where $\rho^{\mathcal{T}}_\alpha$ is the sup of the generators of $E^{\mathcal{T}}_\alpha$) or, symmetrically, $E^{\mathcal{U}}_\alpha$ is the trivial completion of $E^{\mathcal{T}}_\beta \restriction \rho^{\mathcal{U}}_\alpha$. We may as well assume the former. This is a contradiction as in the proof of the comparison lemma unless $[0,\beta]_U \cap D^{\mathcal{U}} = \varnothing$ and $E^{\mathcal{U}}_\beta = \dot{F}^{\mathcal{Q}_\beta}$; that is, $E^{\mathcal{U}}_\beta$ is the unique extender from the $\mathcal{Q}_\beta$ sequence for which we don't have the initial segment condition. But then $\mathcal{Q}_\beta$ is a psuedo-premouse, and thus obeys the initial segment condition on $\dot{F}^{\mathcal{Q}_\beta}$ somewhere past its largest cardinal. It follows that $\rho^{\mathcal{T}}_\alpha \geq$ largest cardinal of $\mathcal{Q}_\beta$. Thus $\mathrm{lh}\, E^{\mathcal{T}}_\alpha$ is not a cardinal of $\mathcal{Q}_\beta$. On the other hand, $\mathrm{lh}\, E^{\mathcal{T}}_\alpha$ is a cardinal of $\mathcal{P}_{\alpha+1}$, hence of $\mathcal{P}_\beta$. This contradicts the fact that $\dot{F}^{\mathcal{Q}_\beta}$ is part of the least disagreement between $\mathcal{P}_\beta$ and $\mathcal{Q}_\beta$, and hence the comparison must terminate.

So let θ be such that $\mathcal{P}_\theta$ is an initial segment of $\mathcal{Q}_\theta$ or vice-versa. The Dodd-Jensen lemma, adapted to our present situation, implies that $\mathcal{P}_\theta = \mathcal{Q}_\theta$, $D^{\mathcal{T}} \cap [0,\theta]_T = \varnothing = D^{\mathcal{U}} \cap [0,\theta]_U$, and $i^{\mathcal{T}}_{0\theta} = i^{\mathcal{U}}_{0\theta}$. The trees involving the extender G must have well founded branches since they can be embedding into trees using F instead of G. Thus we can apply the Dodd-Jensen lemma to a tree involving G even though G is not a member of $\mathcal{M}$.

Now let α be least such that $\alpha + 1 \in (0,\theta)_T$, and β be least such that $\beta + 1 \in (0,\theta)_U$. As $i^{\mathcal{T}}_{0\theta} = i^{\mathcal{U}}_{0,\theta}$ we have that $E^{\mathcal{T}}_\alpha$ and $E^{\mathcal{U}}_\beta$ are compatible up to $\inf(\rho^{\mathcal{T}}_\alpha, \rho^{\mathcal{U}}_\beta)$, that is, either $E^{\mathcal{T}}_\alpha$ is the trivial completion of $E^{\mathcal{U}}_\beta \restriction \rho^{\mathcal{T}}_\alpha$ or $E^{\mathcal{U}}_\beta$ is the trivial completion of $E^{\mathcal{T}}_\alpha \restriction \rho^{\mathcal{U}}_\beta$. When we proved that this comparison terminates we derived a contradiction from this situation using the assumption that both of the ordinals α and β are greater than 0. It follows that at least one of the ordinals α and β must be equal to 0, but on the other hand α and β cannot both be 0, for $E^{\mathcal{T}}_\alpha$ is always G while if $E^{\mathcal{U}}_\beta$ exists then it is equal to $\dot{E}^{\mathcal{M}}_\rho$. If $E^{\mathcal{U}}_\beta$ exists then ρ is a limit ordinal and ρ is a limit of generators of G, but not a limit of generators of $\dot{E}^{\mathcal{M}}_\rho$ and hence G and $\dot{E}^{\mathcal{M}}_\rho$ are not compatible with each other.

Thus we have two cases, depending on which of α and β is equal to 0. Suppose first $\beta = 0$, so that $\alpha \neq 0$ and $E^{\mathcal{U}}_\beta = \dot{E}^{\mathcal{M}}_\rho$ (and $\rho \in \operatorname{dom} \dot{E}^{\mathcal{M}}$). Then, ρ is a limit ordinal, hence a limit of generators of F, and hence a cardinal of $\mathcal{P}_1 = \mathrm{Ult}(\mathcal{M}, G)$. So ρ is a cardinal of $\mathcal{P}_\alpha$. As $\mathcal{P}_\alpha$ is a psuedo-premouse or a premouse, $E^{\mathcal{T}}_\alpha$ satisfies the initial segment condition in $\mathcal{P}_\alpha$ somewhere past ρ. But $\dot{E}^{\mathcal{M}}_\rho$ is the trivial completion of $E^{\mathcal{T}}_\alpha \restriction \nu$, where $\nu < \rho \leq \rho^{\mathcal{T}}_\alpha$ is the sup of the generators of $\dot{E}^{\mathcal{M}}_\rho$. Thus $\dot{E}^{\mathcal{M}}_\rho \in \mathcal{P}_\alpha$, so that ρ is not a cardinal of $\mathcal{P}_\alpha$. This is a contradiction, and hence $\beta \neq 0$.

Now suppose $\alpha = 0$, so that $E^{\mathcal{T}}_\alpha = G$.

First suppose that $\rho - 1$ exists. Then $\rho \notin \operatorname{dom} \dot{E}^{\mathcal{M}}$ so $\mathcal{Q}_1 = \mathcal{Q}_0$. Also, letting $\gamma = \operatorname{lh} G$, $\mathrm{Ult}_0(\mathcal{M}, G)$ and $\mathrm{Ult}_0(\mathcal{M}, F)$ agree below γ, so that $\mathcal{M}$ and $\mathrm{Ult}_0(\mathcal{M}, G)$ agree below γ. That is, $\mathcal{P}_1$ and $\mathcal{Q}_1$ agree below γ. Since γ is a cardinal of $\mathcal{P}_1$, γ is a cardinal of $\mathcal{P}_\beta$. As $\mathcal{Q}_\beta$ is either a premouse or a psuedo-premouse, and $E^{\mathcal{U}}_\beta$ is part of the least disagreement between $\mathcal{P}_\beta$ and $\mathcal{Q}_\beta$ (so that γ is a cardinal in $J^{\mathcal{Q}_\beta}_\eta$, where $\eta = \operatorname{lh} E^{\mathcal{U}}_\beta$), $E^{\mathcal{U}}_\beta \restriction \gamma = G$ is on the sequence of $\mathcal{Q}_\beta$. (One must also consider the $\eta = \gamma$ case. Then $E^{\mathcal{U}}_\beta = G$, and we have the same contradiction.) Thus G is on the sequence of $\mathcal{Q}_1 = \mathcal{Q}_0 = \mathcal{M}$. This contradicts our choice of G, so that ρ is not a successor ordinal.

Suppose next ρ is a limit ordinal, but $\rho \notin \operatorname{dom} \dot{E}^{\mathcal{M}}$. If ρ is not itself a generator of F, then again $\mathrm{Ult}_0(\mathcal{M}, G)$ agrees with $\mathrm{Ult}_0(\mathcal{M}, F)$ below $\gamma = \operatorname{lh} G$, and the argument from the last paragraph yields a contradiction. If ρ is a generator of F, then the natural embedding $\pi : \mathrm{Ult}(\mathcal{M}, G) \rightarrow \mathrm{Ult}(\mathcal{M}, F)$ has critical point ρ, so the agreement is not obvious. Nevertheless, Theorem 8.2 easily implies that $\mathrm{Ult}_0(\mathcal{M}, G)$ and $\mathrm{Ult}_0(\mathcal{M}, F)$ do agree below $\gamma = \operatorname{lh} G = (\rho^+)^{\mathrm{Ult}_0(\mathcal{M}, G)}$. So again we reach a contradiction as in the last paragraph.

Finally, suppose ρ is a limit ordinal and $\rho \in \operatorname{dom} \dot{E}^{\mathcal{M}}$. From Theorem 8.2 we get that $\mathrm{Ult}_0(\mathcal{M}, G)$ agrees with $\mathrm{Ult}_0(\mathrm{Ult}_0(\mathcal{M}, F), \dot{E}^{\mathcal{M}}_\rho)$ below $\gamma = \operatorname{lh} G$, which implies $\mathcal{P}_1$ agrees with $\mathcal{Q}_1 = \mathrm{Ult}_0(\mathcal{M}, \dot{E}^{\mathcal{M}}_\rho)$ below γ. As γ is a cardinal of $\mathcal{P}_1$, γ is a cardinal of $\mathcal{P}_\beta$, hence of $J^{\mathcal{Q}_\beta}_\eta$ where $\eta = \operatorname{lh} E^{\mathcal{U}}_\beta$. Since $\rho \notin \operatorname{dom} \dot{E}^{\mathcal{Q}_\beta}$, and $\mathcal{Q}_\beta$ satisfies at worst the initial segment condition on psuedo-premice, G is on the $\mathcal{Q}_\beta$ sequence. So G is on the sequence of $\mathcal{Q}_1 = \mathrm{Ult}_0(\mathcal{M}, \dot{E}^{\mathcal{M}}_\rho)$. This contradicts our choice of G. □

§11. The Construction

At last we are in a position to construct our extender sequence $\vec{E}$. We will construct the sequence $\vec{E}$ inside of V_θ where θ is least such that $L(V_\theta)$ satisfies that θ is Woodin. Note that every bounded subset of θ in $L(\vec{E})$ is in $L_\theta[\vec{E}]$ since θ is inaccessible.

The construction of $\vec{E}$ will differ from that for sequences of measures in that we do not simply define E_α by induction on α. The reason is that we want the construction to provide each E_α with an ancestry tracing back (by inverting certain collapses) to an extender on V having a certain amount of strength. The illustrious ancestry of the extenders which lie on $\vec{E}$ guarantees that all levels of $L[\vec{E}]$ are ω-iterable.

Let us call a premouse $\mathcal{M}$ *reliable* iff for all $k \leq \omega$, $\mathfrak{C}_k(\mathcal{M})$ exists and is k-iterable. We shall simply assume in this section that the premice we produce in our construction are reliable, and discharge our obligation to show this in §12.

We now define by induction on ξ a reliable coremouse $\mathcal{M}_\xi$. Simultaneously, we verify an induction hypothesis A_ξ describing the agreement between $\mathcal{M}_\xi$ and the $\mathcal{M}_\alpha$ for $\alpha < \xi$:

(A_ξ) $\mathcal{J}_\eta^{\mathcal{M}_\alpha} = \mathcal{J}_\eta^{\mathcal{M}_\xi}$ for all $\alpha < \xi$ and $\kappa \leq \inf\{\rho_\omega(\mathcal{M}_\nu) : \alpha < \nu \leq \xi\}$, where $\eta = (\kappa^+)^{\mathcal{M}_\alpha}$.

In the formulation of A_ξ, we understand that $\omega\eta = \mathrm{OR}^{\mathcal{M}_\alpha}$ in the case that $\mathcal{M}_\alpha \models \kappa^+$ doesn't exist.

We begin by setting $\mathcal{M}_0 = (V_\omega, \in, \varnothing)$. Now suppose that $\mathcal{M}_\xi$ is given and that A_ξ holds. We define $\mathcal{M}_{\xi+1}$ and verify $A_{\xi+1}$.

Case 1. $\mathcal{M}_\xi = (J_\alpha^{\vec{E}}, \varepsilon, \vec{E})$ is a passive premouse, and there are an extender F^* over V, an extender F over $\mathcal{M}_\xi$, and an ordinal $\nu < \alpha$ such that

$$V_{\nu+\omega} \subseteq \mathrm{Ult}(V, F^*)$$

and

$$F \restriction \nu = F^* \cap ([\nu]^{<\omega} \times J_\alpha^{\vec{E}})$$

and

$$\mathcal{N}_{\xi+1} = (J_\alpha^{\vec{E}}, \in, \vec{E}, \tilde{F})$$

is a 1-small, reliable premouse, with $\nu = \nu^{\mathcal{N}_{\xi+1}}$.

In this case we choose F^*, F, ν, and $\mathcal{N}_{\xi+1}$ as above with ν, the natural length of F, minimal among all such F^*. Let

$$\mathcal{M}_{\xi+1} = \mathfrak{C}_\omega(\mathcal{N}_{\xi+1}).$$

Case 2. Otherwise.

In this case, let $\omega\alpha = \mathrm{OR}^{\mathcal{M}_\xi}$, and set

$$\mathcal{N}_{\xi+1} = \left(J_{\alpha+1}^{\dot{E}^{\mathcal{M}_\xi} \frown \dot{F}^{\mathcal{M}_\xi}}, \in, \dot{E}^{\mathcal{M}_\xi} \frown \dot{F}^{\mathcal{M}_\xi}\right).$$

(Of course, $\dot{F}^{\mathcal{M}_\xi} = \varnothing$ is possible.) Thus $\mathcal{N}_{\xi+1}$ is a passive premouse. If $\mathcal{N}_{\xi+1}$ is not reliable, stop the construction. Otherwise,

$$\mathcal{M}_{\xi+1} = \mathfrak{C}_\omega(\mathcal{N}_{\xi+1}).$$

We must verify $A_{\xi+1}$. Now Theorem 8.1 tells us that $\mathcal{N}_{\xi+1}$ agrees with $\mathcal{M}_{\xi+1} = \mathfrak{C}_\omega(\mathcal{N}_{\xi+1})$ below $(\rho_\omega^+)^{\mathcal{N}_{\xi+1}} = (\rho_\omega^+)^{\mathcal{M}_{\xi+1}}$. The obvious agreement between $\mathcal{M}_\xi$ and $\mathcal{N}_{\xi+1}$, together with our induction hypothesis A_ξ, easily gives $A_{\xi+1}$.

Now suppose λ is a limit ordinal. Let

$$\eta = \liminf_{\xi \to \lambda} (\rho_\omega^+)^{\mathcal{M}_\xi}$$

(where again we set $(\rho_\omega^+)^{\mathcal{M}_\xi}$ = unique α s.t. $\omega\alpha = \mathrm{OR}^{\mathcal{M}_\xi}$ in case $\mathcal{M}_\xi \models \rho_\omega^{\mathcal{M}_\xi}$ has no successor cardinal, or $\rho_\omega^{\mathcal{M}_\xi} = \mathrm{OR}^{\mathcal{M}_\xi}$.) Then we let $\mathcal{N}_\lambda$ be the passive premouse $\mathcal{P} = \mathcal{J}_\eta^{\mathcal{P}}$, where for all $\beta < \eta$ we set $\mathcal{J}_\beta^{\mathcal{P}}$ equal to the eventual value of $\mathcal{J}_\beta^{\mathcal{M}_\xi}$ as $\xi \to \lambda$.

$\mathcal{N}_\lambda$ exists since A_ξ holds for all $\xi < \lambda$. Now suppose $\mathcal{N}_\lambda$ is reliable; if not we stop the construction. Set

$$\mathcal{M}_\lambda = \mathfrak{C}_\omega(\mathcal{N}_\lambda).$$

It is easy, using 8.1 and the induction hypothesis, to verify A_λ.

This completes the inductive definition of the $\mathcal{M}_\xi$'s. For the moment, let us assume:

Lemma 11.1. *The construction above never stops; $\mathcal{M}_\xi$ is defined for all ordinals ξ.*

Promise of Proof. We have to show $\mathcal{N}_\xi$ is reliable for all ξ. We will prove as theorem 12.1 that $\mathfrak{C}_k(\mathcal{N}_\xi)$ is k-iterable, for all $k \leq \omega$, provided that $\mathfrak{C}_k(\mathcal{N}_\xi)$ exists. Given this it follows from theorem 8.1 that $\mathcal{N}_\xi$ is reliable. □

Lemma 11.2. *Suppose α_0 and ξ are ordinals such that $\alpha_0 < \xi$ and $\kappa = \rho_\omega^{\mathcal{M}_\xi} \leq \rho_\omega^{\mathcal{M}_\alpha}$ for all $\alpha \geq \alpha_0$. Then $\mathcal{M}_\xi$ is an initial segment of $\mathcal{M}_\eta$, for all $\eta \geq \xi$. Moreover, $\mathcal{M}_{\xi+1} \models$ every set has cardinality at most κ.*

Proof. We may assume $\kappa < \mathrm{OR}^{\mathcal{M}_\xi}$. We claim $\mathcal{M}_{\xi+1}$ is defined by Case 2. For suppose not; let $\mathcal{M}_\xi = (J_\alpha^{\vec{E}}, \in, \vec{E})$ and let F be as in Case 1. Then $\mathcal{M}_\xi$ is a proper initial segment of $\mathrm{Ult}_0(\mathcal{M}_\xi, F)$, and $\mathrm{Ult}_0(\mathcal{M}_\xi, F) \models \alpha$ is a cardinal. As there is a map from κ onto α which is $\Sigma_n^{\mathcal{M}_\xi}$ for some n, we have a contradiction.

Let $\mathcal{M}_\xi = (J_\alpha^{\vec{E}}, \in, \vec{E}, \tilde{F})$, where $F = \varnothing$ if $\mathcal{M}_\xi$ is passive. Let $\mathcal{N}_{\xi+1} = (J_{\alpha+1}^{\vec{E}^\frown F}, \in, \vec{E}^\frown F)$ be as in Case 2. Then the $\Sigma_n^{\mathcal{M}_\xi}$ map from κ onto α guarantees $\mathcal{N}_{\xi+1} \models$ every set has card $= \kappa$, and $\rho_1^{\mathcal{N}_{\xi+1}} \leq \kappa$. Thus $\rho_\omega^{\mathcal{N}_{\xi+1}} = \kappa$. Theorem 8.1 implies that $\mathfrak{C}_\omega(\mathcal{N}_{\xi+1}) = \mathcal{N}_{\xi+1}$. Thus $\mathcal{M}_{\xi+1} = \mathcal{N}_{\xi+1}$, and the claim holds for $\eta = \xi + 1$. For $\eta > \xi + 1$, the claim follows easily from the induction hypothesis A_η. □

The claim implies that $\liminf_{\xi \to \mathrm{OR}} \rho_\omega^{\mathcal{M}_\xi} = \mathrm{OR}$. So we can define our desired $\vec{E}$ by

$$\mathcal{J}_\beta^{\vec{E}} = \text{eventual value of } \mathcal{J}_\beta^{\mathcal{M}_\xi}, \text{ all sufficiently large } \xi \in \mathrm{OR}.$$

Clearly this determines $\vec{E}$, and we have that every level $\mathcal{J}_\beta^{\vec{E}}$ of $L[\vec{E}]$ is an ω-sound, ω-iterable 1-small mouse.

We can think of the construction as producing, in increasing order, the cardinals of $L[\vec{E}]$ together with the levels of $L[\vec{E}]$ whose ωth projectum is a cardinal of $L[\vec{E}]$. Namely, let

$$\kappa_0 = \omega, \quad \xi_0 = 1,$$

and now suppose we have κ_γ and ξ_γ for $\gamma < \alpha$. Set

$$\kappa_\alpha = \inf \{\rho_\omega^{\mathcal{M}_\beta} \mid \beta \geq \sup \{\xi_\gamma \mid \gamma < \alpha\}\}$$

and

$$\xi_\alpha = \text{least } \beta \geq \sup \{\xi_\gamma \mid \gamma < \alpha\} \text{ such that } \rho_\omega^{\mathcal{M}_\beta} = \kappa_\alpha .$$

One can check easily that $\langle \kappa_\alpha \mid \alpha \in \mathrm{OR}\rangle$ enumerates in non-decreasing order the cardinals of $L[\vec{E}]$, that $\rho_\omega^{\mathcal{M}_{\xi_\alpha}} = \kappa_\alpha$, and that $\mathcal{M}_{\xi_\alpha}$ is a level of the eventual $L[\vec{E}]$. In fact, for κ a cardinal of $L[\vec{E}]$, the $\mathcal{M}_{\xi_\alpha}$ for $\kappa_\alpha = \kappa$ are precisely those levels $\mathcal{J}_\beta^{\vec{E}}$ of $L[\vec{E}]$ whose ωth projectum is κ.

We now show that $L[\vec{E}] \models$ there is a Woodin cardinal. Once again, certain iterability assumptions will crop up during the proof. We shall verify these assumptions in §12.

Theorem 11.3. *Suppose there is a Woodin cardinal. Let $\vec{E}$ be the extender sequence constructed above. Then $L[\vec{E}] \models$ there is a Woodin cardinal.*

PROOF. Let θ be least such that $L(V_\theta) \models$ "θ is Woodin." We show that θ is Woodin in $L[\vec{E}]$.

So fix $f : \theta \to \theta$ such that $f \in L[\vec{E}]$. Define $g : \theta \to \theta$ in V by

$$g(\alpha) = \text{2nd strongly inaccessible (of } V) > f(\alpha).$$

(The strong inaccessible is just a security blanket.) As θ is Woodin in V there is an extender F^* over V, $F^* \in V_\theta$, crit $F^* = \kappa$, such that if $j^* : V \to \mathrm{Ult}(V, F^*)$ is the canonical embedding, then $g''\kappa \subseteq \kappa$ and

$$V_{j^*(g)(\kappa)+1} \subseteq \mathrm{Ult}(V, F^*)$$

and

$$\vec{E} \restriction j^*(g)(\kappa) = j^*(\vec{E}) \restriction j^*(g)(\kappa).$$

Let

$$F = F^* \cap ([\mathrm{lh}\, F^*]^{<\omega} \times L[\vec{E}]).$$

Notice that $L[\vec{E}]$ agrees with $\mathrm{Ult}(L[\vec{E}], F)$ below $j^*(g)(\kappa)$, where the ultrapower is computed using functions in $L[\vec{E}]$. That is, F "coheres" with $\vec{E}$ sequence out to $j^*(g)(\kappa)$. Notice $j^*(g)(\kappa)$ is a strongly inaccessible cardinal of V, hence of $L[\vec{E}]$. We now show that for $\rho < j^*(g)(\kappa)$, the trivial completion of $F \restriction \rho$ is on $\vec{E}$, or an ultrapower thereof.

Let $(\kappa^+)^{L[\vec{E}]} \le \rho < j^+(g)(\kappa)$. Let

$$i : L[\vec{E}] \to \mathrm{Ult}(L[\vec{E}], F \restriction \rho)$$

be the canonical embedding, and let

$$\gamma = (\rho^+)^{\mathrm{Ult}(L[\vec{E}], F\restriction\rho)},$$
$$G = \{(a, x) \mid a \in [\gamma]^{<\omega} \wedge x \subseteq [\kappa]^{\mathrm{card}(a)} \wedge x \in L[\vec{E}] \wedge a \in i(x)\}.$$

Thus G is the trivial completion of $F \restriction \rho$. The generators of G are of course just those generators of F which are less than ρ, and $G \restriction \rho = F \restriction \rho$.

Lemma 11.4. *Let $(\kappa^+)^{L[\vec{E}]} \le \rho < j^*(g)(\kappa)$, and suppose that ρ is the natural length of $F \restriction \rho$. Let G be the trivial completion of $F \restriction \rho$, and $\gamma = \mathrm{lh}\, G$. Then $E_\gamma = G = F \restriction \rho$ unless ρ is a limit ordinal greater than $(\kappa^+)^{L[\vec{E}]}$, and is itself a generator of F. In this case*

$$G = \begin{cases} E_\gamma & \text{if } \gamma \notin \mathrm{dom}\, \vec{E} \\ (i^{E_\rho}(\vec{E}))_\gamma & \text{if } \gamma \in \mathrm{dom}\, \vec{E} \end{cases},$$

where $i^{E_\rho} : J_\rho^{\vec{E}} \to \mathrm{Ult}_0(J_\rho^{\vec{E}}, E_\rho)$ is the canonical embedding.

Proof (modulo §12). The proof proceeds by induction on ρ, and is divided into a number of cases. In those cases where ρ is not a cardinal we will apply theorem 10.1, and in the other cases we will be able to use bicephali.

Case A. ρ is a successor. In this case $\rho - 1$ must be a generator of F. Let

$$\sigma : \mathrm{Ult}(L[\vec{E}], F \restriction \rho) \to \mathrm{Ult}(L[E], F)$$

be the canonical embedding. From the case hypothesis we see that $\sigma(\rho) = \rho$ and hence $\sigma \restriction \gamma = \mathrm{id}$. Also, $G = F \restriction \gamma$ as $(a, x) \in G \Leftrightarrow a \in i_{F\restriction\rho}(x) \Leftrightarrow a \in \sigma(i_{F\restriction\rho}(x)) \Leftrightarrow a \in i_F(x) \Leftrightarrow (a, x) \in F$, for all $a \in [\gamma]^{<\omega}$ and appropriate x.

We claim there is a stage η of the construction such that

$$\mathcal{M}_\eta = (J_\gamma^{\vec{E}}, \in, \vec{E} \restriction \gamma).$$

For let δ be the largest cardinal of $L[\vec{E}]$ which is $\leq \rho$. (So $\delta < \rho$.) Now $i_{F\restriction\rho}(\vec{E}) \restriction \gamma = i_F(\vec{E}) \restriction \gamma = \vec{E} \restriction \gamma$, and by the definition of γ, we have $J_\gamma^{i_{F\restriction\rho}(\vec{E})} \models$ every set has cardinality $\leq \rho$, so $J_\gamma^{\vec{E}} \models$ every set has cardinality $\leq \rho$. On the other hand, δ is the largest cardinal $\leq \rho$ in $L[\vec{E}]$, hence in $J_{\sigma(\gamma)}^{i_F(\vec{E})} = J_{\sigma(\gamma)}^{\vec{E}}$, (as $\sigma(\gamma)$ is a cardinal of $L[\vec{E}]$) hence in $J_\gamma^{\vec{E}}$. So

$$J_\gamma^{\vec{E}} \models \text{every set has cardinality } \delta\,.$$

Let $\langle \xi_\alpha \mid \alpha < (\delta^+)^{L[\vec{E}]} \rangle$ enumerate in increasing order those ordinals ξ such that $\rho_\omega(\mathcal{M}_\xi) = \delta$ and $\rho_\omega(\mathcal{M}_\beta) \geq \delta$ for all $\beta \geq \xi$. We observed earlier that the $\mathcal{M}_{\xi\alpha}$ are precisely those levels of $L[\vec{E}]$ whose ωth projectum is δ. It is clear that γ is a limit of such levels. So letting $\eta = \sup \{\xi_\alpha \mid \mathrm{OR} \cap \mathcal{M}_{\xi_\alpha} < \gamma\}$, we have that η is a limit and $\mathcal{M}_\eta = (J_\gamma^{\vec{E}}, \in, \vec{E} \restriction \gamma)$. This proves our claim.

Now clearly $(J_\gamma^{\vec{E}}, \in, \vec{E} \restriction \gamma, \tilde{F} \restriction \gamma)$ is a type II premouse. It is also 1-small, since otherwise $J_\kappa^{\vec{E}}$ satisfies that some ordinal $\alpha < \kappa$ is Woodin. But then since κ is a cardinal of $L[\vec{E}]$, α is Woodin in $L[\vec{E}]$, and $\alpha < \kappa < \theta$, contrary to our initial assumption that no ordinal $\alpha < \theta$ is Woodin in $L[\vec{E}]$.

Let us assume until §12:

Sublemma 11.4.1. $(J_\gamma^{\vec{E}}, \in, \vec{E} \restriction \gamma, \tilde{F} \restriction \gamma)$ *is reliable.*

It follows that $\mathcal{M}_{\eta+1}$ is defined by Case 1 in our construction. That is, $\mathcal{N}_{\eta+1} = (J_\gamma^{\vec{E}}, \in, \vec{E} \restriction \gamma, H)$ for some H, and $\mathcal{M}_{\eta+1} = \mathfrak{C}_\omega(\mathcal{N}_{\eta+1})$. But $\rho_\omega(\mathcal{N}_{\eta+1}) \geq \delta$ since δ is the largest cardinal of $L[\vec{E}]$ and hence $\mathcal{M}_{\eta+1} = \mathcal{N}_{\eta+1}$ since $J_\gamma^{\vec{E}}$ satisfies that every set has cardinality at most δ. Moreover $\mathcal{M}_{\eta+1}$ is an initial segment of the eventual $L[\vec{E}]$, and $H = E_\gamma$. Thus it is enough to show $E_\gamma = F \restriction \gamma$.

Notice that γ is a generator of F, as otherwise $\sigma(\gamma) = \gamma$, so that γ is a cardinal of $L[\vec{E}]$, contrary to $\gamma \in \mathrm{dom}\, \vec{E}$. Let G' be the trivial completion of $F \restriction \gamma + 1$. Arguing as above, with $\xi = \mathrm{lh}\, G'$, we see that $G' = F \restriction \xi$, and that $\vec{E} \restriction \xi^\frown F \restriction \xi$ satisfies conditions 1-4 of "good at ξ". We now show that $(J_\xi^{\vec{E}}, \in, \vec{E} \restriction \xi, F \restriction \xi)$ is a psuedo-premouse.

Since it is easy to see that δ remains the largest cardinal of $J_\xi^{\vec{E}}$, as γ is not a cardinal of $J_\xi^{\vec{E}}$, we need to verify that the trivial completion of $F \restriction \delta$ is on $\vec{E}$. Now either $\delta = (\kappa^+)^{L[\vec{E}]}$ or δ is a limit of generators of F. [Otherwise, let $\tau < \delta$ be such that $\tau = \bigcup\{\xi < \delta \mid \xi = (\kappa^+)^{L[\vec{E}]}$ or ξ is a generator of $F\}$. By our inductive hypothesis the trivial completion of $F \restriction \tau$ is on $\vec{E}$ - it falls under either (b) or (e) of the lemma. But from $F \restriction \tau$ we easily construct a collapse of δ.] By our inductive hypothesis, as $\delta < \rho$, the trivial completion of $F \restriction \delta$ is on $\vec{E}$. (Note here that clause (d) of the lemma cannot apply as $\delta \notin \operatorname{dom} \vec{E}$ as δ is a cardinal of $L[\vec{E}]$.) That is, if $\beta = (\delta^+)^{\mathrm{Ult}(L[\vec{E}], F \restriction \delta)}$, then E_β is the trivial completion of $F \restriction \delta$. Clearly $\beta \leq \gamma < \xi$. Thus $(J_\xi^{\vec{E}}, \in, \vec{E} \restriction \xi, F \restriction \xi)$ satisfies the initial segment condition on psuedo-premice, as desired.

We now borrow from §12:

Sublemma 11.4.2. $(J_\xi^{\vec{E}}, \in, \vec{E} \restriction \xi, F \restriction \xi)$ *is iterable.*

Granted 11.4.2, Theorem 10.1 tells us that $(J_\xi^{\vec{E}}, \in, \vec{E} \restriction \xi, F \restriction \xi)$ satisfies the full initial segment condition, so that $F \restriction \gamma = E_\gamma$.

Remark. We can't use bicephali here because E_γ might be of type III, while $F \restriction \gamma$ is of type II.

Case B. ρ is a limit of generators of F, but not itself a generator of F.

Let $\sigma : \mathrm{Ult}(L[\vec{E}], F \restriction \rho) \to \mathrm{Ult}(L[\vec{E}], F)$ by the canonical embedding. As ρ is not a generator of F, $\sigma \restriction \gamma = \mathrm{id}$ and $G = F \restriction \gamma$. Note ρ is a cardinal of $J_\gamma^{\vec{E}}$, hence of $L[\vec{E}]$ because σ exists.

Arguing exactly as in Case A we find a stage η of the construction such that

$$\mathcal{M}_\eta = (J_\gamma^{\vec{E}}, \in, \vec{E} \restriction \gamma)$$

and

$$(J_\gamma^{\vec{E}}, \in, \vec{E} \restriction \gamma, F \restriction \gamma)$$

is a premouse of type III. In §12 we prove:

Sublemma 11.4.3. $(J_\gamma^{\vec{E}}, \in, \vec{E} \restriction \gamma, F \restriction \gamma)$ *is reliable.*

Thus $\mathcal{M}_{\eta+1}$ is defined through Case 1 of our construction. Let H be the set such that $\mathcal{N}_{\eta+1} = (J_\gamma^{\vec{E}}, \in, \vec{E} \restriction \gamma, H)$. Now ρ is the largest cardinal of $(J_\gamma^{\vec{E}}, \in, \vec{E} \restriction \gamma)$, and we chose H so as to minimize $\nu^{\mathcal{N}_{\eta+1}}$, the sup of the generators of H. Thus $\nu^{\mathcal{N}_{\eta+1}} = \rho$ and $\mathcal{N}_{\eta+1}$ is of type III or type I. Drawing on §12, we get

Sublemma 11.4.4. *The structure* $(J_\gamma^{\vec{E}}, \in, \vec{E} \restriction \gamma, F \restriction \gamma, H)$ *is an iterable type III bicephalus.*

It follows from Theorem 9.2 that $H = F \restriction \gamma$. Also, as in Case A, $\mathcal{M}_{\eta+1} = \mathfrak{C}_\omega(\mathcal{N}_\eta) = \mathcal{N}_\eta$, and $\mathcal{M}_{\eta+1}$ (note here $\rho = \rho_\omega(\mathcal{N}_\eta)$ is the largest cardinal of $\mathcal{N}_\eta$, and ρ is a cardinal of $L[\vec{E}]$) is an initial segment of $L[\vec{E}]$. Thus $\gamma \in \operatorname{dom} \vec{E}$ and $E_\gamma = F \restriction \gamma$.

Case C. ρ is a limit of generators of F, and is itself a generator of F, and $\rho \notin \operatorname{dom} \vec{E}$.

Again, let

$$\sigma : \operatorname{Ult}(L[\vec{E}], F \restriction \rho) \to \operatorname{Ult}(L[\vec{E}], F)$$

be the canonical embedding. This time we have $\rho = \operatorname{crit} \sigma$, and thus it is not obvious that G "coheres" with $\vec{E}$ up to γ. Nevertheless, Theorem 8.2 implies that this is true.

Claim 1. $\operatorname{Ult}(L[\vec{E}], F \restriction \rho)$ agrees with $L[\vec{E}]$ below γ.

Proof. Let η be any ordinal such that $\rho < \eta < \gamma$ and $\rho_\omega^{\mathcal{H}} = \rho$ where $\mathcal{H} = \mathcal{J}_\eta^{\operatorname{Ult}(L[\vec{E}], F \restriction \rho)}$. Since γ is the successor cardinal of ρ in $\operatorname{Ult}(L[\vec{E}], F \restriction \rho)$, there are arbitrarily large such ordinals $\eta < \gamma$. It will thus be enough to see that $\mathcal{H}$ is an initial segment of $L[\vec{E}]$. But now $\sigma \restriction \mathcal{H}$ is a fully elementary map from $\mathcal{H}$ into $\sigma(\mathcal{H})$; moreover $\operatorname{crit}(\sigma \restriction \mathcal{H}) = \rho_\omega^{\mathcal{H}}$ and $\rho_\omega^{\mathcal{H}} \notin \operatorname{dom} \vec{E}$ (so $\rho_\omega^{\mathcal{H}} \notin \operatorname{dom} \dot{E}^{\sigma(\mathcal{H})}$). Thus Theorem 8.2 implies that $\mathcal{H}$ is an initial segment of $\sigma(\mathcal{H})$. But $\sigma(\mathcal{H})$ agrees with $L[\vec{E}]$ below $\operatorname{lh} F$, hence below γ, and thus $\mathcal{H}$ is an initial segment of $L[\vec{E}]$.

Claim 2. $(J_\gamma^{\vec{E}}, \in, \vec{E} \restriction \gamma, G)$ is a 1-small type III premouse.

Proof. For coherence, we use Claim 1. The initial segment condition follows from our induction hypothesis on ρ. We get 1-smallness as in Case A.

We now consider two subcases.

Subcase C 1. ρ is a cardinal of $L[\vec{E}]$. In this case we have, just as in Case A, that there is a stage η of our construction such that

$$\mathcal{M}_\eta = (J_\gamma^{\vec{E}}, \in, \vec{E} \restriction \gamma).$$

(Here η is the sup of all ξ_α s.t. $\kappa_\alpha = \rho$ and $\rho < \mathrm{OR}^{\mathcal{M}_{\xi_\alpha}} < \gamma$.) Granted this, we will proceed just as in Case B:

Sublemma 11.4.5. $(J_\gamma^{\vec{E}}, \in, \vec{E} \restriction \gamma, G)$ *is reliable.*

PROOF. In §12.

So $\mathcal{M}_{\eta+1}$ is defined via Case 1 in our construction. Let

$$\mathcal{N}_{\eta+1} = (J_\gamma^{\vec{E}}, \in, \vec{E} \restriction \gamma, H).$$

As we chose H to minimize $\nu^{\mathcal{N}_{\eta+1}}$, $\mathcal{N}_{\eta+1}$ is either type I or type III, and $\nu^{\mathcal{N}_{\eta+1}} = \rho$.

Sublemma 11.4.6. *The structure* $(J_\gamma^{\vec{E}}, \in, \vec{E} \restriction \gamma, G, H)$ *is an iterable type* III *bicephalus.*

Proof. In §12.

From Theorem 9.2 we get that $G = H$. As in Case B, Theorem 8.1 and Lemma 11.2 guarantee that $\mathcal{M}_{\eta+1} = \mathcal{N}_{\eta+1}$ and $\mathcal{M}_{\eta+1}$ is an initial segment of $L[\vec{E}]$. Thus $G = E_\gamma$.

Subcase C 2. ρ is not a cardinal of $L[\vec{E}]$. We use the argument from Case A. Let δ be the largest cardinal of $L[\vec{E}]$ which is $< \rho$. Let G' be the trivial completion of $F \restriction \rho + 1$, and $\xi = \mathrm{lh}\, G'$. Thus $G' = F \restriction \xi$, and $(J_\xi^{\vec{E}}, \in, \vec{E} \restriction \xi, \tilde{F} \restriction \xi)$ satisfies conditions 1 through 4 of goodness at ξ. Using σ we see that δ is the largest cardinal of $J_\gamma^{\vec{E}}$ which is less than ρ. It follows that δ is the largest cardinal of $J_\xi^{\vec{E}}$ (note ρ is not a cardinal in $J_\xi^{\vec{E}}$ since the natural embedding from $\mathrm{Ult}(L[\vec{E}], F \restriction \rho + 1)$ into $\mathrm{Ult}(L[\vec{E}], F)$ fixes ρ, and ρ is not a cardinal of $\mathrm{Ult}(L[\vec{E}], F)$). Our induction hypothesis guarantees that the trivial completion of $F \restriction \delta$ is on $\vec{E}$, and hence on $\vec{E} \restriction \xi$. Thus $(J_\xi^{\vec{E}}, \in, \vec{E} \restriction \xi, \tilde{F} \restriction \xi)$ is a psuedo-premouse.

Sublemma 11.4.7. $(J_\xi^{\vec{E}}, \in, \vec{E} \restriction \xi, \tilde{F} \restriction \xi)$ *is iterable.*

Proof. In §12.

Theorem 10.1 implies that $(J_\xi^{\vec{E}}, \in, \vec{E} \restriction \xi, \tilde{F} \restriction \xi)$ satisfies the full initial segment condition on premice, so that $\gamma \in \mathrm{dom}\, \vec{E}$ and $E_\gamma = F \restriction \gamma$, as desired.

Remark. We do not seem to get that there is a stage η of the construction such that $\mathcal{M}_\eta = (J_\gamma^{\vec{E}}, \in, \vec{E} \restriction \gamma)$ in Subcase C2.

Case D. ρ is a limit of generators of F, a generator of F itself, and $\rho \in \mathrm{dom}\, \vec{E}$.

As $\rho \in \mathrm{dom}\, \vec{E}$, ρ is not a cardinal of $L[\vec{E}]$. We can now just repeat the argument from Subcase C2. Letting ξ be the length of the trivial completion of $F \restriction \rho + 1$, $(J_\xi^{\vec{E}}, \in, \vec{E} \restriction \xi, \tilde{F} \restriction \xi)$ is a psuedo-premouse and, borrowing from §12, is iterable. By Theorem 10.1, $(J_\xi^{\vec{E}}, \in, \vec{E} \restriction \xi, \tilde{F} \restriction \xi)$ satisfies the full initial segment condition on premice. As $\rho \in \mathrm{dom}\, \vec{E}$, this means G is on the sequence of $\mathrm{Ult}((J_\rho^{\vec{E}}, \in, E \restriction \rho), E_\rho)$, as desired.

Case E. $\rho = (\kappa^+)^{L[\vec{E}]}$.

The proof is the same as that in Case B. We omit further detail.

This completes the proof of Lemma 11.4. □

We can now easily finish the proof of Theorem 11.3. Let

$$\rho = \text{ least strongly inaccessible cardinal of } L[\vec{E}] > j^*(f)(\kappa).$$

Let G be the trivial completion of $F \restriction \rho$, and $\gamma = \operatorname{lh} G$. By the choice of j^* we know that ρ is definable in $\mathrm{Ult}(L[\vec{E}], F)$ from $j^*(f)(\kappa) < \rho$ and hence is not a generator of F. Thus lemma 11.4 implies that $\gamma \in \operatorname{dom} \vec{E}$ and $E_\gamma = F \restriction \gamma$. We have the diagram

$$\begin{array}{ccc} L[\vec{E}] & \xrightarrow{\;j^*\;} & \mathrm{Ult}(L[\vec{E}], F^*) \\ & \searrow^{\,i} & \uparrow k \\ & & \mathrm{Ult}(L[\vec{E}], F \restriction \gamma) \end{array}$$

where the upper ultrapower is computed using functions in V, and the lower using functions in $L[\vec{E}]$. The function k is defined by $k([a,h]^{L[\vec{E}]}_{F\restriction\gamma}) = [a,h]^V_{F^*}$. Since $k \restriction \gamma = \mathrm{id}$, $i(f)(\kappa) < \rho$. By coherence, $J^{\vec{E}}_\rho = J^{\mathrm{Ult}(L[\vec{E}],F\restriction\gamma)}_\rho$, and thus

$$L[\vec{E}] \models V_\rho \subseteq \mathrm{Ult}(L[\vec{E}], F \restriction \gamma).$$

So $F \restriction \gamma$ witnesses the Woodin property for the function f. □

§12. ITERABILITY

We now discharge our obligation to show that various of the structures we encountered in §11 are iterable. We shall concentrate on proving Lemma 11.1, which states, in the language of §11, that $\mathcal{N}_\eta$ is reliable for all η. The other iterability lemmas from §11 are proved in almost the same way. A complete proof of these lemmas will be given in the paper [S?a].

As we observed in §11, it is enough to show

Theorem 12.1. *Let $\mathcal{N}_\eta$ be the ηth "$\mathcal{N}$-model" of the construction of §11. Let $0 \leq k \leq \omega$ and suppose $\mathfrak{C}_k(\mathcal{N}_\eta)$ exists. Then $\mathfrak{C}_k(\mathcal{N}_\eta)$ is k-iterable.*

PROOF. The proof of theorem 12.1 will take up all of this final section of the paper. Let

$$\mathcal{T} = \langle T, \deg, D, \langle E_\alpha^{\mathcal{T}}, \mathcal{P}_{\alpha+1}^* \mid \alpha + 1 < \theta \rangle \rangle$$

be a k-bounded, k-maximal iteration tree on

$$\mathcal{P}_0 = \mathfrak{C}_k(\mathcal{N}_\eta) \, .$$

The assumption of k-maximality is not necessary, but it simplifies the notation a bit, and we have never used non-maximal trees anyway. We let $\mathcal{P}_\alpha$ be the αth model of $\mathcal{T}$. Suppose that $\mathcal{T} \restriction \lambda$ is simple for all $\lambda < \theta$, and that θ is a limit ordinal. We shall show that $\mathcal{T}$ has a cofinal wellfounded branch.

For $\gamma < \eta$ such that Case 1 applied in the definition of $\mathcal{M}_{\gamma+1}$ from $\mathcal{M}_\gamma$, that is, such that $\mathcal{N}_{\gamma+1}$ is equal to $\mathcal{M}_\gamma$ expanded by a new predicate for a last extender, we let F_γ^* be the background extender for the last extender of $\mathcal{N}_{\gamma+1}$. Thus F_γ^* is $\nu + \omega$ strong, where $\nu = \nu^{\mathcal{N}_{\gamma+1}}$. Set

$$\mathbb{C} = (\langle \mathcal{N}_\gamma \mid \gamma \leq \eta \rangle, \langle F_\gamma^* \mid \gamma < \eta \text{ and } F_\gamma^* \text{ defined} \rangle) \, .$$

Our strategy for the proof of theorem 12.1 is straightforward. We shall associate to $\mathcal{T}$ a tree $\mathcal{U}$ which will be an iteration tree on V in the sense of [MS]. As such the models of $\mathcal{U}$ will be well founded by results methods of [MS]. The tree ordering of $\mathcal{U}$ will be the same tree ordering, T, as $\mathcal{T}$, and we will define embeddings π_α from the models of $\mathcal{U}$ to those of $\mathcal{U}$. Thus the models of the tree $\mathcal{T}$ will also be well founded, which is what we need to show.

Since $\mathcal{U}$ is not a fine structure iteration tree it doesn't make sense to ask that $\vec{\pi}$ be a tree-embedding in the sense of section 5. However, if we let R_α be the αth model of $\mathcal{U}$ then the embeddings π_α will be embeddings from the αth model $\mathcal{P}_\alpha$ of $\mathcal{T}$ into $Q_\alpha = \mathfrak{C}_j(\mathcal{S}) \in R_\alpha$ where $\mathcal{S}$ is on the sequence of models of $i_0^{\mathcal{U}}(\mathbb{C})$ and $j = \deg(\alpha)$. If we modify the definition of a tree-embedding for this case by asking that π_α be a $(\deg^{\mathcal{T}}, Y_\alpha)$-embedding into Q_α instead of into R_α then $\vec{\pi}$ will satisfy this definition.

We must also maintain a certain amount of agreement between π_α and the π_β's for $\beta < \alpha$. We now state some definitions which allow us to describe this agreement.

DEFINITION 12.1.1. Let $\mathcal{M}$ be a premouse, and $\omega\lambda \leq \mathrm{OR}^{\mathcal{M}}$. Then the *$\lambda$-dropdown sequence of* $\mathcal{M}$ is the sequence $\langle\langle\beta_0, k_0\rangle, \ldots, \langle\beta_i, k_i\rangle\rangle$ defined as follows:

(1) $\langle\beta_0, k_0\rangle = \langle\lambda, 0\rangle$.
(2) $\langle\beta_{i+1}, k_{i+1}\rangle$ is the lexicographically least pair $\langle\beta, k\rangle$ such that $\lambda \leq \beta$, $\omega\beta \leq \mathrm{OR}^{\mathcal{M}}$, and $\rho_k(\mathcal{J}_\beta^{\mathcal{M}}) < \rho_{k_i}(\mathcal{J}_{\beta_i}^{\mathcal{M}})$.

If there is no such pair $\langle\beta, k\rangle$ then $\langle\beta_{i+1}, k_{i+1}\rangle$ is undefined. Let i be the largest integer such that $\langle\beta_i, k_i\rangle$ is defined.

Notice that if $\langle\langle\beta_e, k_e\rangle \mid e \leq i\rangle$ is the λ-dropdown sequence of $\mathcal{M}$, then $k_e < \omega$ for all $e \leq i$ and

$$\langle\beta_e, k_e\rangle <_{\mathrm{lex}} \langle\beta_{e+1}, k_{e+1}\rangle$$

for all $e < i$. Moreover every ordinal of the form $\rho = \rho_k(\mathcal{J}_\beta^{\mathcal{M}})$ for $k \in \omega$, $\beta\omega \leq \mathrm{OR}^{\mathcal{M}}$, and $\rho \leq \lambda \leq \beta$ is in the set $\{\,\rho_{k_e}(\mathcal{J}_{\beta_e}^{\mathcal{M}}) \mid e \leq i\,\}$.

Now we prepare to define the *(j,ξ)-resurrection sequence* for an extender E, where E is on the extender sequence of $\mathcal{M} = \mathfrak{C}_j(\mathcal{N}_\xi)$, the jth core of one of the models of our construction $\mathbb{C}$. We are allowing the possibility that $E = \dot{F}^{\mathcal{M}}$. The idea is just to trace E back to its origin as the last extender of some $\mathcal{N}_\gamma$ with $\gamma \leq \xi$.

Let $\lambda = \mathrm{lh}\, E$, and suppose that $\langle\langle\beta_0, k_0\rangle \cdots \langle\beta_i, k_i\rangle\rangle$ is the initial segment of the λ-dropdown sequence of $\mathcal{M}$ consisting of those pairs $\langle\beta, k\rangle$ on the sequence such that $\langle\beta, k\rangle \leq_{\mathrm{lex}} \langle\alpha, j\rangle$, where $\omega\alpha = \mathrm{OR}^{\mathcal{M}}$. Our first goal is to show that there is a unique $\gamma \leq \xi$ such that $\mathcal{J}_{\beta_i}^{\mathcal{M}} = \mathfrak{C}_{k_i}(\mathcal{N}_\gamma)$. Fix α such that $\omega\alpha = \mathrm{OR}^{\mathcal{M}}$ and let $\kappa = \rho_{k_i}(\mathcal{J}_{\beta_i}^{\mathcal{M}})$.

CLAIM 1. Let $\langle\gamma, e\rangle \leq_{\mathrm{lex}} \langle\xi, j\rangle$ and suppose $\mathcal{J}_{\beta_i}^{\mathcal{M}}$ is an initial segment of $\mathfrak{C}_e(\mathcal{N}_\gamma)$. Then for all $\langle\tau, n\rangle$ such that $\langle\gamma, e\rangle \leq_{\mathrm{lex}} \langle\tau, n\rangle \leq_{\mathrm{lex}} \langle\xi, j\rangle$, $\mathcal{J}_{\beta_i}^{\mathcal{M}}$ is an initial segment of $\mathfrak{C}_n(\mathcal{M}_\tau)$.

PROOF. Let $\kappa = \rho_{k_i}(\mathcal{J}_{\beta_i}^{\mathcal{M}})$, which is the minimum value of $\rho_k(\mathcal{J}_\beta^{\mathcal{M}})$ for pairs $\langle\beta, k\rangle$ satisfying $\langle\lambda, 0\rangle \leq_{\mathrm{lex}} \langle\beta, k\rangle \leq_{\mathrm{lex}} \langle\alpha, j\rangle$. Notice that $\mathcal{J}_{\beta_i}^{\mathcal{M}}$ is k_i-sound, since $k_i \leq j$ if $\beta_i = \alpha$. It will suffice then to show that $\rho_n(\mathcal{N}_\tau) \geq \kappa$ whenever $\langle\gamma, e\rangle \leq_{\mathrm{lex}} \langle\tau, n\rangle \leq_{\mathrm{lex}} \langle\xi, j\rangle$. (We leave the details here to the reader.) So suppose $\mu < \kappa$ and $\mu = \rho_n(\mathcal{N}_\tau)$ for some such $\langle\tau, n\rangle$. Let μ be the minimal value of such a $\rho_n(\mathcal{N}_\tau)$. Then $\mathfrak{C}_n(\mathcal{N}_\tau)$ is an initial segment of $\mathfrak{C}_j(\mathcal{N}_\xi) = \mathcal{M}$ by 8.1. The minimality of κ implies $\mathfrak{C}_n(\mathcal{N}_\tau)$ is a proper initial segment of $\mathcal{J}_\lambda^{\mathcal{M}}$. This contradicts that there is a subset of μ which is definable over $\mathfrak{C}_n(\mathcal{N}_\tau)$ but not a member of $J_\kappa^{\mathcal{M}}$. □

CLAIM 2. If $\langle\gamma, e{+}1\rangle \leq_{lex} \langle\xi, j\rangle$ and $\mathcal{J}_{\beta i}^{\mathcal{M}}$ is a proper initial segment of $\mathfrak{C}_{e+1}(\mathcal{N}_\gamma)$, then $\mathcal{J}^{\mathcal{M}}_{\beta_i}$ is a proper initial segment of $\mathfrak{C}_e(\mathcal{N}_\gamma)$.

PROOF. By the 1st claim $\rho_{e+1}(\mathcal{N}_\gamma) \geq \kappa$. But $\beta_i < (\kappa^+)^{\mathfrak{C}_{e+1}(\mathcal{N}_\gamma)}$ since $\mathcal{J}^{\mathcal{M}}_{\beta_i}$ has projectum κ. By 8.1, $\mathcal{J}^{\mathcal{M}}_{\beta_i}$ is a proper initial segment of $\mathfrak{C}_e(\mathcal{N}_\gamma)$. □

CLAIM 3. There is a unique $\gamma \leq \xi$ s.t. $\mathcal{J}^{\mathcal{M}}_{\beta_i} = \mathfrak{C}_{k_i}(\mathcal{N}_\gamma)$.

PROOF. Let $\langle \gamma, e \rangle$ be $\leq_{\text{lex}}$ least such that $\mathcal{J}^{\mathcal{M}}_{\beta_i}$ is an initial segment of $\mathfrak{C}_e(\mathcal{N}_\gamma)$.

Suppose first that $\mathcal{J}^{\mathcal{M}}_{\beta_i} = \mathfrak{C}_e(\mathcal{N}_\gamma)$. If $e \leq k_i$, then since $\mathcal{J}^{\mathcal{M}}_{\beta_i}$ is k_i sound, $\mathcal{J}^{\mathcal{M}}_{\beta_i} = \mathfrak{C}_{k_i}(\mathcal{N}_\gamma)$ as desired. To see that $e \leq k_i$, suppose toward a contradiction that $k_i < e$. For t, $u \leq e$ set

$$\rho^u_t = \rho_t(\mathfrak{C}_u(\mathcal{N}_\gamma)).$$

It will be enough to see that $\rho^{k_i}_{k_i} = \rho^e_e$, since this implies that $\mathfrak{C}_{k_i}(\mathcal{N}_\gamma) = \mathfrak{C}_e(\mathcal{N}_\gamma) = \mathcal{J}^{\mathcal{M}}_{\beta_i}$, contrary to the minimality of $\langle \gamma, e \rangle$. So suppose we have t s.t. $k_i \leq t < e$ and $\rho^{t+1}_{t+1} < \rho^t_t$. We may assume t is the largest such. Now the reader can easily check[1] that for any u, $\rho^u_{u+1} = \rho^{u+1}_{u+1}$, and $\rho^u_{u+1} < \rho^u_u \Rightarrow \rho^{u+1}_{u+1} < \rho^{u+1}_u$. Thus we have $\rho^{t+1}_{t+1} < \rho^{t+1}_t \leq \rho^{t+1}_{k_i}$. As $\mathfrak{C}_{t+1}(\mathcal{N}_\gamma) = \mathfrak{C}_e(\mathcal{N}_\gamma)$ by the maximality of t, $\rho^e_e = \rho^{t+1}_{t+1} < \rho^{t+1}_{k_i} = \rho^e_{k_i}$. But $\mathfrak{C}_e(\mathcal{N}_\gamma) = \mathcal{J}^{\mathcal{M}}_{\beta_i}$, so this contradicts the fact that $\langle \beta_i, k_i \rangle$ is the last term of this restriction of the λ-dropdown sequence of $\mathcal{M}$, so that $\rho_e(\mathcal{J}^{\mathcal{M}}_{\beta_i}) < \rho_{k_i}(\mathcal{J}^{\mathcal{M}}_{\beta_i})$ is impossible if $k_i < e \leq j$. If $\mathcal{J}^{\mathcal{M}}_{\beta_i} = \mathcal{M}$ then we must verify that $e \leq j$ in order to apply this fact. Now if $\gamma = \xi$, then $e \leq j$ by the choice of $\langle \gamma, e \rangle$, and if $\gamma < \xi$, then $\mathfrak{C}_e(\mathcal{N}_\gamma) = \mathfrak{C}_j(\mathcal{N}_\xi)$ and it is easy to see that this is impossible.

Next, suppose $\mathcal{J}^{\mathcal{M}}_{\beta_i}$ is a proper initial segment of $\mathfrak{C}_e(\mathcal{N}_\gamma)$. From Claim 2, we see that $e = 0$, so that $\mathcal{J}^{\mathcal{M}}_{\beta_i}$ is a proper initial segment of $\mathcal{N}_\gamma$.

If γ is a limit, then the definition of $\mathcal{N}_\gamma$ guarantees that $\mathcal{J}^{\mathcal{M}}_{\beta_i}$ is a proper initial segment of some $\mathfrak{C}_\omega(\mathcal{N}_\tau)$ for $\tau < \gamma$. But then Claim 2 implies $\mathcal{J}^{\mathcal{M}}_{\beta_i}$ is a proper initial segment of $\mathcal{N}_\tau$, a contradiction. Thus γ is a successor.

Let $\gamma = \tau + 1$. From the definition of $\mathcal{N}_{\tau+1}$ (either we add an extender predicate to $\mathcal{M}_\tau$ or extend the J-hierarchy for one more step), $\mathcal{J}^{\mathcal{M}}_{\beta_i}$ is an initial segment of $\mathcal{M}_\tau = \mathfrak{C}_\omega(\mathcal{N}_\tau)$. This contradicts the minimality of $\langle \gamma, e \rangle$.

Thus $\mathcal{J}^{\mathcal{M}}_{\beta_i} = \mathfrak{C}_{k_i}(\mathcal{N}_\gamma)$ for some $\gamma \leq \xi$. There is a unique such γ by the following easy fact, whose proof we omit: if $\gamma \neq \delta$ then $\mathfrak{C}_e(\mathcal{N}_\gamma) \neq \mathfrak{C}_k(\mathcal{N}_\delta)$, $\forall \gamma, \delta, e, k$. □

We can now define the (j, ξ) *resurrection sequence for* E.

CASE 1. $i = 0$. Notice that $\rho_1(\mathcal{J}^{\mathcal{M}}_\lambda) < \lambda$, since $\mathcal{J}^{\mathcal{M}}_\lambda$ is active. Since $\langle \beta_1, k_1 \rangle = \langle \lambda, 1 \rangle$ is not defined we must have $\mathcal{N}_\xi = \mathcal{M} = \mathcal{J}^{\mathcal{M}}_\lambda$ and $j = 0$. Then E is the

[1]By the way, it is also true, though not at all obvious, that $\rho^u_u < \rho^u_{u-1} \Rightarrow \rho^{u+1}_u < \rho^{u+1}_{u-1}$.

last extender of $\mathcal{N}_\xi$, and the (j,ξ) resurrection sequence for E is defined to be the empty sequence.

CASE 2. $i > 0$. Let $\gamma \leq \xi$ be such that $\mathcal{J}_{\beta_i}^{\mathcal{M}} = \mathfrak{C}_{k_i}(\mathcal{N}_\gamma)$. Notice that $k_i \geq 1$ as $\rho_{k_i}(\mathcal{N}_\gamma) < \lambda$ and $\omega\lambda \leq \mathrm{OR}^{\mathcal{N}_\gamma}$. Let $\pi : \mathfrak{C}_{k_i}(\mathcal{N}_\gamma) \to \mathfrak{C}_{k_i-1}(\mathcal{N}_\gamma)$ be the inverse of the collapse. Then the (j,ξ) resurrection sequence for E is $\langle \beta_i, k_i, \gamma, \pi\rangle^\frown S$, where S is the $(k_i - 1, \gamma)$ resurrection sequence for $\pi(E)$. (Here if E is the last extender of $\mathfrak{C}_{k_i}(\mathcal{N}_\gamma)$, then by $\pi(E)$ we mean the last extender of $\mathfrak{C}_{k_i-1}(\mathcal{N}_\gamma)$.)

This completes the recursive definition of the (j,ξ) resurrection sequence for E.

For any premouse $\mathcal{P}$ with $\omega\alpha = \mathrm{OR}^{\mathcal{P}}$ and $t < \omega$, and $\omega\lambda \leq \mathrm{OR}^{\mathcal{P}}$, the (t,λ) *dropdown sequence* of $\mathcal{P}$ is just that initial segment of the λ-dropdown sequence of $\mathcal{P}$ consisting of pairs (β, k) such that $(\beta, k) \leq_{\mathrm{lex}} (\alpha, t)$.

Now let us return to the situation of Case 2 of the definition of the (j,ξ) resurrection sequence for E, and adopt the notation there. Let us adopt our standard notational device by taking $\pi(\mathrm{OR}^{\mathfrak{C}_{k_i}(\mathcal{N}_\gamma)})$ to be $\mathrm{OR}^{\mathfrak{C}_{k_i-1}(\mathcal{N}_\gamma)}$. One can easily see from our results on preservation of projecta that the $(k_i - 1, \pi(\lambda))$ dropdown sequence for $\mathfrak{C}_{k_i-1}(\mathcal{N}_\gamma)$, which is what we use to resurrect $\pi(E)$, has the form

$$\langle\langle \pi(\beta_0), k_0\rangle, \ldots, \langle \pi(\beta_{i-1}), k_{i-1}\rangle\rangle^\frown u\,,$$

where

$$u = \varnothing \quad \text{or} \quad u = \langle \pi(\beta_i), k_i - 1\rangle\,.$$

We do not know whether it is possible that $u \neq \varnothing$. In order for this to happen we would need to have $\langle \beta_{i-1}, k_{i-1}\rangle \neq \langle \beta_i, k_i - 1\rangle$, $\rho_{k_i-1}(\mathcal{J}_{\beta_i}^{\mathcal{M}}) = \rho_{k_{i-1}}(\mathcal{J}_{\beta_{i-1}}^{\mathcal{M}})$, and $\rho_{k_i-1}(\mathfrak{C}_{k_i-1}(\mathcal{N}_\gamma)) < \pi(\rho_{k_i-1}(\mathcal{J}_{\beta_i}^{\mathcal{M}}))$. We only know that $\rho_{k_i-1}(\mathfrak{C}_{k_i-1}(\mathcal{N}_\gamma)) = \sup \pi''\rho_{k_i-1}(\mathcal{J}_{\beta_i}^{\mathcal{M}})$. It seems plausible that π preserves the $k_i - 1$st projectum, so that in fact $u = \varnothing$ must hold.

Notice that if $u \neq \varnothing$, then the last integer k_i in the dropdown sequence gets decreased by 1 at the next stage of resurrection. Thus there are cofinally many stages in the resurrection at which the u associated to the stage is $\varnothing$. These stages are important, so we now give a formal definition.

Let E be on the sequence of $\mathfrak{C}_j(\mathcal{N}_\xi)$, $\lambda = \mathrm{lh}\, E$, and let

$$\langle\langle \beta_0, k_0\rangle, \ldots, \langle \beta_i, k_i\rangle\rangle$$

be the (j,λ) dropdown sequence of $\mathfrak{C}_j(\mathcal{N}_\xi)$, and let

$$\langle\langle \delta_0, \ell_0, \gamma_0, \pi_0\rangle, \ldots, \langle \delta_t, \ell_t, \gamma_t, \pi_t\rangle\rangle$$

be the (j,ξ) resurrection sequence for E. (We suppose the resurrection to be non-empty. Thus $(\beta_1, k_1) = (\lambda, 1)$ is defined.) We have at once from the definitions that

$$(\delta_0, \ell_0) = (\beta_i, k_i),$$
$$\mathcal{J}_{\delta_0}^{\mathfrak{C}_j(\mathcal{N}_\xi)} = \mathfrak{C}_{\ell_0}(\mathcal{N}_{\gamma_0}),$$
$$\pi_0 : \mathfrak{C}_{\ell_0}(\mathcal{N}_{\gamma_0}) \to \mathfrak{C}_{\ell_0 - 1}(\mathcal{N}_{\gamma_0}),$$

and for $1 \leq e \leq t$,

$$(\delta_e, \ell_e) = \text{last term in the } (\ell_{e-1} - 1, \pi_{e-1} \circ \cdots \circ \pi_0(\lambda)) \text{ dropdown sequence of } \mathfrak{C}_{\ell_{e-1}-1}(\mathcal{N}_{\gamma_{e-1}}),$$

$$\mathcal{J}_{\delta_e}^{\mathfrak{C}_{\ell_{e-1}-1}(\mathcal{N}_{\gamma_{e-1}})} = \mathfrak{C}_{\ell_e}(\mathcal{N}_{\gamma_e}),$$

and

$$\pi_e : \mathfrak{C}_{\ell_e}(\mathcal{N}_{\gamma_e}) \to \mathfrak{C}_{\ell_e - 1}(\mathcal{N}_{\gamma_e}).$$

From our earlier remarks on the new dropdown sequences, we can find stages

$$1 \leq e_1 < e_2 < \cdots < e_{i-1} = t$$

such that

$$\begin{aligned}
(\delta_{e_1}, \ell_{e_1}) &= \pi_{e_1 - 1} \circ \cdots \circ \pi_0((\beta_{i-1}, k_{i-1})) \\
(\delta_{e_2}, \ell_{e_2}) &= \pi_{e_2 - 1} \circ \cdots \circ \pi_0((\beta_{i-2}, k_{i-2})) \\
&\vdots \\
(\delta_{e_{i-1}}, \ell_{e_{i-1}}) &= \pi_{e_{i-1} - 1} \circ \cdots \circ \pi_0((\beta_1, k_1)).
\end{aligned}$$

Here if $e_1 = 1$, the notation "$\pi_{e_1 - 1} \circ \cdots \circ \pi_0$" stands for π_0. We also set $e_0 = 0$, and interpret "$\pi_{e_0 - 1} \circ \cdots \circ \pi_0$" to stand for the identity embedding. We then have for $0 \leq n \leq i - 1$

$$(\delta_{e_n}, \ell_{e_n}) = \pi_{e_n - 1} \circ \cdots \circ \pi_0((\beta_{i-n}, k_{i-n})).$$

This enables us to define embeddings and models resurrecting the various $\mathcal{J}_{\beta_e}^{\mathcal{M}}$, where $\mathcal{M} = \mathfrak{C}_j(\mathcal{N}_\xi)$. Set

$$\sigma_{i-n} = \pi_{e_n} \circ \pi_{e_n - 1} \circ \cdots \circ \pi_1 \circ \pi_0$$

so that

$$\sigma_{i-n} : \mathcal{J}_{\beta_{i-n}}^{\mathcal{M}} \to \mathfrak{C}_{\ell_{e_n} - 1}(\mathcal{N}_{\gamma_{e_n}})$$

is an $\ell_{e_n} - 1$ embedding, for $0 \leq n \leq i - 1$. In order to simplify the indexing a bit set $\tau_{i-n} = \gamma_{e_n}$ for $0 \leq n \leq i - 1$. Notice also that $k_{i-n} = \ell_{e_n}$. Thus, setting $p = i - n$, we have that for $1 \leq p \leq i$

$$\sigma_p : \mathcal{J}_{\beta_p}^{\mathcal{M}} \to \mathfrak{C}_{k_p - 1}(\mathcal{N}_{\tau_p})$$

is a $k_p - 1$ embedding. Let us set

$$\mathrm{Res}_p = \mathfrak{C}_{k_p-1}(\mathcal{N}_{\tau_p})$$

and call $(\sigma_p, \mathrm{Res}_p)$ the pth *partial resurrection of E from stage* (j,ξ). (Notice that if $p < q$, then Res_p represents "more resurrection" than Res_q in the sense that it goes back to an earlier model $\mathcal{N}_\eta$ and hence nearer to the first appearance of the prototype of E. On the other hand, Res_p resurrects less of $\mathcal{M}$ in the sense that the domain $\mathcal{J}^{\mathcal{M}}_{\beta_p}$ of σ_p is smaller than that of σ_q.

The partial resurrections of E agree with one another in the following way: For $1 \leq p \leq i$, let

$$\kappa_p = \rho_{k_p}(\mathcal{J}^{\mathcal{M}}_{\beta_p}).$$

Then one can check without too much difficulty that $\kappa_1 > \kappa_2 > \cdots > \kappa_i$, and that if $p < q$ then $\sigma_p \restriction \kappa_{q-1} = \sigma_q \restriction \kappa_{q-1}$ and the models Res_p and Res_q agree below $\sup \sigma_q'' \kappa_{q-1}$. For example, consider the case $q = i$. Then $\sigma_i = \pi_0 : \mathcal{J}^{\mathcal{M}}_{\beta_i} \to \mathfrak{C}_{k_i-1}(\mathcal{N}_{\gamma_0})$, and moreover, the last term of the $(k_i - 1, \pi_0(\lambda))$ dropdown sequence for $\mathfrak{C}_{k_1-1}(\mathcal{N}_{\gamma_0})$ corresponds to a projectum which is greater than or equal to $\sup(\pi_0'' \kappa_{i-1})$. This implies that $\pi_j \restriction \sup \pi_0'' \kappa_{i-1}$ is the identity, for all $j > 0$. So

$$\sigma_p \restriction \kappa_{i-1} = \pi_{e_{i-p}} \circ \cdots \circ \pi_1 \circ \pi_0 \restriction \kappa_{i-1} = \pi_0 \restriction \kappa_{i-1} = \sigma_i \restriction \kappa_{i-1}.$$

See figure 1 for a diagram of some of the relationships above.

Finally, the *complete resurrection* of E from (j,ξ) is the pair (identity, $\mathcal{N}_\xi$) if the (j,ξ) resurrection sequence for E is $\emptyset$ (so that $j = 0$ and E is the last extender of $\mathcal{N}_\xi$), and the pair $(\sigma_1, \mathrm{Res}_1)$ if the (j,ξ) resurrection sequence for E is nonempty.

Notice that in any case, $\mathrm{Res} = \mathcal{N}_\gamma$ for some $\gamma \leq \xi$ and σ is a 0-embedding from $\mathcal{J}^{\mathfrak{C}_j(\mathcal{N}_\xi)}_\lambda$ into $\mathcal{N}_\gamma$.

Of course, the notions associated to resurrection can be interpreted not just in $V = R_0$, but in any model R_α of the tree $\mathcal{U}$ (using the construction $i^{\mathcal{U}}_{0\alpha}(\mathbb{C})$). We shall do this in what follows.

Definition of $\mathcal{U}$: Induction hypotheses. During the recursive definition of the tree $\mathcal{U}$ and the embeddings π_α we will be maintaining a number of induction hypotheses, which we have numbered H1 through H7. Recall that R_α is the αth model of the tree $\mathcal{U}$.

H1. There is an ordinal ξ such that the map π_α is a weak n-embedding from $\mathcal{P}_\alpha$ into Q_α, where $n = \deg^{\mathcal{T}} \alpha$ and $Q_\alpha = (\mathfrak{C}_n(\mathcal{N}_\xi))^{R_\alpha}$.

H2. (commutativity) If $\beta T \alpha$ and $(\beta, \alpha]_T \cap D^{\mathcal{T}} = \emptyset$ then $\pi_\alpha \circ i^{\mathcal{T}}_{\beta,\alpha} = i^{\mathcal{U}}_{\beta,\alpha} \circ \pi_\beta$.

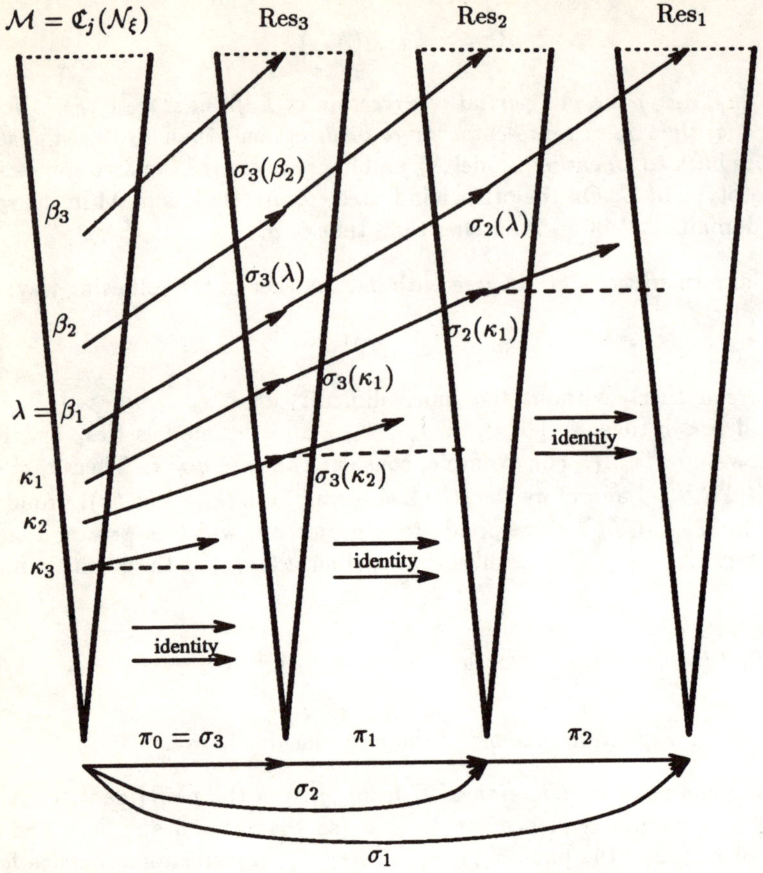

FIGURE 1. To simplify matters, this diagram assumes that $i = 3$ and $\beta_1 < \beta_2 < \beta_3$. It also assumes that the new dropdown sequence is just the image of the old minus its last term, that is, that "$u = \varnothing$" always holds. Thus $t = 2$, $e_0 = 0$, $e_1 = 1$, and $e_2 = 2$. Also, $\sigma_3 = \pi_0$, $\sigma_2 = \pi_1 \circ \pi_0$, and $\sigma_1 = \pi_2 \circ \pi_1 \circ \pi_0$. Finally, we assume that $\pi_e(\rho_{\ell_e - 1}(\mathfrak{C}_{\ell_e}(\mathcal{N}_{\gamma_e}))) = \rho_{\ell_e - 1}(\mathfrak{C}_{\ell_e - 1}(\mathcal{N}_{\gamma_e}))$ for $e = 1, 2$, which, together with a similar assumption on π_0, implies $u = \varnothing$.

Next, we have some agreement of models and embeddings to maintain. For each ordinal $\beta < \mathrm{lh}\, \mathcal{T}$, let ν_β be the natural length of $E_\beta^{\mathcal{T}}$ and let $(\sigma^\beta, \mathrm{Res}^\beta)$ be the complete resurrection of $\pi_\beta(E_\beta^{\mathcal{T}})$ from stage (j, τ), where $j = \deg^{\mathcal{T}}(\beta)$ and $Q_\beta = (\mathfrak{C}_j(\mathcal{N}_\tau))^{R_\beta}$.

H3. For each $\beta < \alpha$, if Res^β is type I or III then Q_α agrees with Res^β below

ν^{Res^β}, moreover

$$\pi_\alpha \restriction \nu_\beta = \sigma^\beta \circ \pi_\beta \restriction \nu_\beta \quad \text{and} \quad \pi_\alpha(\nu_\beta) \geq \nu^{\mathrm{Res}^\beta}.$$

H4. For each $\beta < \alpha$, if Res^β is type II then Q_α agrees with Res^β below $\mathrm{OR}^{\mathrm{Res}^\beta}$, and moreover

$$\pi_\alpha \restriction \mathrm{lh}\, E^{\mathcal{T}}_\beta = \sigma^\beta \circ \pi_\beta \restriction \mathrm{lh}\, E^{\mathcal{T}}_\beta \quad \text{and} \quad \pi_\alpha(\mathrm{lh}\, E^{\mathcal{T}}_\beta) \geq \mathrm{OR}^{\mathrm{Res}^\beta}.$$

H5. For each $\beta < \alpha$, R_α agrees with R_β below $\nu^{\mathrm{Res}^\beta} + \omega$, that is $V_\gamma^{R_\alpha} = V_\gamma^{R_\beta}$ where $\gamma = \nu^{\mathrm{Res}^\beta} + \omega$.

In order to handle the limit case in the definition of $\mathcal{U}$, we will require two final induction hypotheses.

If $Q = \mathfrak{C}_k(\mathcal{N}_\gamma)$ and $Q' = \mathfrak{C}_j(\mathcal{N}_\xi)$ where $\mathcal{N}_\gamma$ and $\mathcal{N}_\xi$ are two models of the construction $\mathbb{C}$, then we write $Q \leq_{\mathbb{C}} Q'$ iff $(\gamma, k) \leq_{\mathrm{lex}} (\xi, j)$.

H6. Let $\beta = \mathcal{T}\text{-Pred}(\alpha+1)$ and $\mathbb{C}^{\alpha+1} = i^{\mathcal{U}}_{0,\alpha+1}(\mathbb{C})$. Then

(a) $Q_{\alpha+1} \leq_{\mathbb{C}^{\alpha+1}} i^{\mathcal{U}}_{\beta,\alpha+1}(Q_\beta)$, and
(b) if $\alpha + 1 \in D^{\mathcal{T}}$, then $Q_{\alpha+1} <_{\mathbb{C}^{\alpha+1}} i^{\mathcal{U}}_{\beta,\alpha+1}(Q_\beta)$.

H7. If λ is a limit ordinal then $i^{\mathcal{U}}_{\alpha\lambda}(Q_\alpha) = Q_\lambda$ for all sufficiently large $\alpha T \lambda$.

We shall need to know that $\mathcal{U}$ is a tree in the "coarse structure" sense of [MS]. Set $\rho^{\mathcal{U}}_\beta = \nu^{\mathrm{Res}^\beta}$. Then it will be obvious from the construction that $E^{\mathcal{U}}_\beta$ is $\rho^{\mathcal{U}}_\beta + \omega$ strong in the model R_β. We shall show in the remark following claim 1 below that $\rho^{\mathcal{U}}_\beta < \rho^{\mathcal{U}}_\delta$ whenever $\beta < \delta$, and the agreement condition on the models R_β follows at once from this. This guarantees that $\mathcal{U}$ is a normal iteration tree in the sense of [MS], provided that no illfounded model appears in $\mathcal{U}$. Thus we need to know that we encounter no illfounded ultrapowers or direct limits in the formation of $\mathcal{U}$. This follows from the following theorem, which is proved by the methods of [MS].

Theorem. *If there is no ordinal $\gamma \leq \xi$ such that $L(V_\gamma) \models$ "γ is a Woodin cardinal" then every iteration tree on $L(V_\xi)$ has a unique cofinal wellfounded branch.*

Note that if theorem 12.1 holds for all $\eta' < \eta$, so that $\mathcal{N}_\eta$ exists, then $\mathcal{N}_\eta$ is constructed in V_ξ for some ordinal ξ smaller than the least cardinal δ such that $L[V_\delta]$ satisfies that δ is a Woodin cardinal. Thus we can apply this theorem to the trees derived from $\mathcal{U}$.

We now begin the recursive definition of the tree $\mathcal{U}$ and the embeddings π_α. For $\alpha = 0$ we take $Q_0 = \mathcal{P}_0$, $R_0 = L(V_\theta)$ where θ is the least ordinal γ such that $L(V_\gamma) \models$ "γ is a Woodin cardinal", and $\pi_0 = $ identity.

Definition of $\mathcal{U}$: The Successor Step. We assume that the tree has been defined through the αth model R_α, and we have the embeddings π_α mapping P_α into Q_α, where $j = \deg^{\mathcal{T}}(\alpha)$ and $Q_\alpha = (\mathfrak{C}_j(\mathcal{N}_\xi))^{R_\alpha}$, and we have (in R_α)

$$(\sigma^\alpha, \mathrm{Res}^\alpha) = \text{complete resurrection of } \pi_\alpha(E^{\mathcal{T}}_\alpha) \text{ from } (j, \xi),$$

where $\pi_\alpha(E^{\mathcal{T}}_\alpha)$ is the last extender predicate of Q_α in case $E^{\mathcal{T}}_\alpha$ is the last extender predicate of $\mathcal{P}_\alpha$.

CLAIM 1. If γ is strictly smaller than α then $\sigma^\alpha \restriction \pi_\alpha(\mathrm{lh}\, E^{\mathcal{T}}_\gamma) = \text{identity}$.

PROOF. Fix $\gamma < \alpha$. Then $\mathrm{lh}\, E^{\mathcal{T}}_\gamma$ is a cardinal of $\mathcal{P}_\alpha$, so $\pi_\alpha(\mathrm{lh}\, E^{\mathcal{T}}_\gamma)$ is a cardinal of Q_α. Thus $\rho_\omega(\mathcal{J}^{Q_\alpha}_\beta) \geq \pi_\alpha(\mathrm{lh}\, E^{\mathcal{T}}_\gamma)$ for all β such that $\pi_\alpha(\mathrm{lh}\, E^{\mathcal{T}}_\alpha) \leq \omega\beta < \mathrm{OR}^{Q_\alpha}$. We claim that also $\rho_j(Q_\alpha) \geq \pi_\alpha(\mathrm{lh}\, E^{\mathcal{T}}_\gamma)$. (Recall that $j = \deg(\alpha)$.) Assume first that α is a successor ordinal. Then $\mathcal{P}_\alpha = \mathrm{Ult}_j(\mathcal{P}^*_\alpha, E^{\mathcal{T}}_{\alpha-1})$, and so $\mathrm{lh}\, E^{\mathcal{T}}_{\alpha-1} < \rho_j(\mathcal{P}_\alpha)$. Thus $\mathrm{lh}\, E^{\mathcal{T}}_\gamma < \rho_j(\mathcal{P}_\alpha)$, and as π_α is a weak j-embedding, $\pi_\alpha(\mathrm{lh}\, E^{\mathcal{T}}_\gamma) < \rho_j(Q_\alpha)$. Now our claim for the case α is a limit ordinal follows from the successor case applied to sufficiently large $\alpha' T \alpha$.

Thus no projectum associated to a term in the $(j, \pi_\alpha(\mathrm{lh}\, E^{\mathcal{T}}_\alpha))$ dropdown sequence for Q_α lies below $\pi_\alpha(\mathrm{lh}\, E^{\mathcal{T}}_\gamma)$, and it follows that σ^α is the identity below $\pi_\alpha(\mathrm{lh}\, E^{\mathcal{T}}_\gamma)$.

REMARK. The claim enables us to show that $\rho^{\mathcal{U}}_\alpha > \rho^{\mathcal{U}}_\beta$ for all $\beta < \alpha$. For

$$\rho^{\mathcal{U}}_\alpha = \nu^{\mathrm{Res}^\alpha} = \sigma^\alpha \circ \pi_\alpha(\nu_\alpha).$$

But now, for $\beta < \alpha$, $\mathrm{lh}\, E^{\mathcal{T}}_\beta$ is a cardinal of $\mathcal{P}_\alpha$ and $\mathrm{lh}\, E^{\mathcal{T}}_\beta < \mathrm{lh}\, E^{\mathcal{T}}_\alpha$, so that $\mathrm{lh}\, E^{\mathcal{T}}_\beta \leq \nu_\alpha$. Thus $\nu_\beta < \nu_\alpha$ for $\beta < \alpha$. So

$$\rho^{\mathcal{U}}_\alpha > \sigma^\alpha \circ \pi_\alpha(\nu_\beta).$$

But Claim 1 tells us $\sigma^\alpha \circ \pi_\alpha(\nu_\beta) = \pi_\alpha(\nu_\beta)$, and our induction hypotheses on agreement of embeddings say $\pi_\alpha(\nu_\beta) \geq v^{\mathrm{Res}^\beta}$. So

$$\rho^{\mathcal{U}}_\alpha > \sigma^\alpha \circ \pi_\alpha(\nu_\beta) = \pi_\alpha(\nu_\beta) \geq \nu^{\mathrm{Res}^\beta} = \rho^{\mathcal{U}}_\beta.$$

We can now define $E^{\mathcal{U}}_\alpha$ and $R_{\alpha+1}$. Set

$$F = \sigma^\alpha \circ \pi_\alpha(E^{\mathcal{T}}_\alpha) = \text{last extender of } \mathrm{Res}^\alpha.$$

Now Res^α is an "$\mathcal{N}$ model" in the universe R_α, so its last extender has a "background extender". Set $E^{\mathcal{U}}_\alpha = F^*$, the background extender for F in R_α. Let $\beta = T\text{-pred}(\alpha + 1)$ and set

$$R_{\alpha+1} = \mathrm{Ult}(R_\beta, F^*).$$

Notice that $\mathrm{Ult}_0 = \mathrm{Ult}_\omega$ since $R_\beta \models ZFC$.

Let us note that R_α and R_β are in sufficient agreement that this ultrapower makes sense. This is clear if $\beta = \alpha$, so we may suppose that $\beta < \alpha$. By our induction hypotheses, R_α agrees with R_β to $\nu^{\mathrm{Res}^\beta} + \omega$. Now crit $E_\alpha^{\mathcal{T}} < \nu_\beta$ because $\beta = T\text{-pred}(\alpha+1)$. As σ^α is the identity on $\pi_\alpha(\mathrm{lh}\, E_\beta^{\mathcal{T}})$, crit $F^* =$ crit $F =$ crit $\sigma^\alpha(\pi_\alpha(E_\alpha^{\mathcal{T}})) < \sup \pi_\alpha'' \nu_\beta = \sup \sigma^\beta \circ \pi_\beta'' \nu_\beta \leq \nu^{\mathrm{Res}^\beta}$. Thus the ultrapower makes sense.

We now define $\pi_{\alpha+1}$ and $Q_{\alpha+1}$. Let $n = \deg^{\mathcal{T}}(\beta)$, and $\lambda = \mathrm{lh}\, E_\beta^{\mathcal{T}}$, let

$$\langle(\eta_0, k_0), \ldots, (\eta_e, k_e)\rangle \text{ be the } (n, \lambda) \text{ dropdown sequence of } \mathcal{P}_\beta,$$

and set $\kappa_i = \rho_{k_i}(\mathcal{J}_{\eta_i}^{\mathcal{P}_\beta})$ for $0 \leq i \leq e$.

The following claim relates these to the $(n, \pi_\beta(\lambda))$ dropdown sequence of Q_β. The claim is slightly complicated by the fact that π_β is not a full n-embedding. Notice that $\kappa_e \leq \rho_n(\mathcal{P}_\beta)$.

Claim 2. The $(n, \pi_\beta(\lambda))$-dropdown sequence of Q_β is the sequence given by the appropriate clause below:

(a) If $\kappa_e < \rho_n(\mathcal{P}_\beta)$ then the dropdown sequence is

$$\langle(\pi_\beta(\eta_0), k_0), \ldots, (\pi_\beta(\eta_e), k_e)\rangle.$$

(b) If $\kappa_e = \rho_n(\mathcal{P}_\beta)$ but $(\omega\eta_e, k_e) \neq (\mathrm{OR}^{\mathcal{P}_\beta}, n)$ then the dropdown sequence is

$$\langle(\pi_\beta(\eta_0), k_0), \ldots, (\pi_\beta(\eta_e), k_e)\rangle^\frown u,$$

where $u = \varnothing$ or $u = (\eta, n)$ for $\omega\eta = \mathrm{OR}^{Q_\beta}$.

(c) If $(\omega\eta_e, k_e) = (\mathrm{OR}^{\mathcal{P}_\beta}, n)$ then the dropdown sequence is

$$\langle(\pi_\beta(\eta_0), k_0), \ldots, (\pi_\beta(\eta_{e-1}), k_{e-1})\rangle^\frown u,$$

where $u = \varnothing$ or $u = (\pi_\beta(\eta_e), k_e) = (\omega\eta, n)$, for $\omega\eta = \mathrm{OR}^{Q_\beta}$.

Remark. Note that $\kappa_e = \rho_n(\mathcal{P}_\beta)$ in case (c). If $e = 0$, then $n = 0 = k_0$ and $\eta_0 = \lambda = \omega\lambda = \mathrm{OR}^{\mathcal{P}_\beta}$. The $(n, \pi_\beta(\lambda))$ dropdown sequence for Q_β is then $\langle(\mathrm{OR}^{Q_\beta}, 0)\rangle$, which falls under case (c).

Remark. The $u = \varnothing$ case in (c) would not be necessary if π_β were a full n-embedding.

The claim follows easily from the fact that π_β is a weak n-embedding. For (a), notice that $\pi_\beta(\kappa_e) < \sup \pi_\beta''\rho_n(\mathcal{P}_\beta) \leq \rho_n(Q_\beta)$. Recall that π_β preserves cardinals, so that if for example $\omega\eta_e < \mathrm{OR}^{\mathcal{P}_\beta}$ then $\mathcal{P}_\beta \models \forall\gamma \geq \eta_e(\rho_\omega(\mathcal{J}_\gamma^{\dot{E}}) \geq \rho_{k_e}(\mathcal{J}_{\eta_e}^{\dot{E}}))$, and thus $Q_\beta \models \forall\gamma \geq \pi_\beta(\eta_e)(\rho_\omega(\mathcal{J}_\gamma^{\dot{E}}) \geq \pi_\beta(\kappa_e))$. □

Let $\mu = \mathrm{crit}(E_\alpha^{\mathcal{T}})$, and let

$$i = \begin{cases} e+1 & \text{if } \mu < \kappa_e, \\ \text{least } j \text{ s.t. } \kappa_j \leq \mu & \text{if } \kappa_e \leq \mu. \end{cases}$$

Notice that $i > 0$ since $\kappa_0 = \lambda > \mu$. Because $\mathcal{T}$ is maximal

$$\mathcal{P}^*_{\alpha+1} = \begin{cases} \mathcal{J}_{\eta_i}^{\mathcal{P}_\beta} & \text{if } i \leq e, \\ \mathcal{P}_\beta & \text{if } i = i+1, \end{cases}$$

and

$$\deg^{\mathcal{T}}(\alpha+1) = \begin{cases} k_i - 1 & \text{if } i \leq e, \\ n & \text{if } i = e+1. \end{cases}$$

Let $(\sigma_i^\beta, \mathrm{Res}_i^\beta)$ be the ith partial resurrection of $\pi_\beta(E_\beta^{\mathcal{T}})$ from stage (n,τ), where $Q_\beta = \mathfrak{C}_n(\mathcal{N}_\tau)^{R_\beta}$, if this resurrection is defined. The resurrection is undefined if $i = e+1$ and defined if $i < e$ by claim 2. If $i = e$ then $(\sigma_i^\beta, \mathrm{Res}_i^\beta)$ is undefined just in case $(\omega\eta_e, k_e) = (\mathrm{OR}^{\mathcal{P}_\beta}, n)$ and the conclusion of (c) of claim 2 holds with $u = \varnothing$.

Now let

$$Q^*_{\alpha+1} = \begin{cases} \mathrm{Res}_i^\beta & \text{if } \mathrm{Res}_i^\beta \text{ is defined} \\ Q_\beta & \text{otherwise,} \end{cases}$$

$$\sigma = \begin{cases} \sigma_i^\beta & \text{if } \mathrm{Res}_i^\beta \text{ is defined,} \\ \text{identity} & \text{otherwise.} \end{cases}$$

Then $\sigma \circ (\pi_\beta \restriction \mathcal{P}^*_{\alpha+1})$ is, in any case, a weak $\deg^{\mathcal{T}}(\alpha+1)$ embedding from $\mathcal{P}^*_{\alpha+1}$ into $Q^*_{\alpha+1}$. To see this, assume first that Res_i^β is defined, so that $i \leq e$, $\deg^{\mathcal{T}}(\alpha+1) = k_i - 1$, and $\sigma = \sigma_i^\beta$ is a full $(k_i - 1)$ embedding. Looking at claim 2, we see that in all cases the domain of σ is $\mathcal{J}_{\pi_\beta(\eta_i)}^{Q_\beta}$ since we cannot have the situation in (c) with $i = e$ and $u = \varnothing$. But $\mathcal{P}^*_{\alpha+1} = \mathcal{J}_{\eta_i}^{\mathcal{P}_\beta}$, and $\pi_\beta \restriction \mathcal{P}^*_{\alpha+1}$ is a weak $(k_i - 1)$ embedding. In fact, if $\omega\eta_i < \mathrm{OR}^{\mathcal{P}_\beta}$ then $\pi_\beta \restriction \mathcal{P}^*_{\alpha+1}$ is fully elementary, and if $\omega\eta_i = \mathrm{OR}^{\mathcal{P}_\beta}$ then $k_i \leq n$, so $\pi_\beta \restriction \mathcal{P}^*_{\alpha+1}$ is a weak k_i-embedding. It follows that $\sigma \circ (\pi_\beta \restriction \mathcal{P}^*_{\alpha+1})$ is a weak $k_i - 1$-embedding from $\mathcal{P}^*_{\alpha+1}$ into $Q^*_{\alpha+1}$. Assume next that Res_i^β is undefined. Then either $i = e+1$ or we have the situation in (c) of claim 2 with $u = \varnothing$. In either case, $\deg^{\mathcal{T}}(\alpha+1) \leq n$. Also $\mathcal{P}^*_{\alpha+1} = \mathcal{P}_\beta$, $Q^*_{\alpha+1} = Q_\beta$, and σ is the identity. Since π_β is a weak n-embedding, $\sigma \circ \pi_\beta$ is a weak $\deg^{\mathcal{T}}(\alpha+1)$-embedding from $\mathcal{P}^*_{\alpha+1}$ into $Q^*_{\alpha+1}$.

Let $Q_\beta = \mathfrak{C}_n(\mathcal{N}_\tau)^{R_\beta}$, so that $(\sigma^\beta, \mathrm{Res}^\beta)$ is the complete resurrection of $\pi_\beta(E_\beta^{\mathcal{T}})$ from stage (n,τ). Let ψ be the complete resurrection embedding for $\sigma \circ \pi_\beta(E_\beta^{\mathcal{T}})$ from the appropriate stage, which is (n,τ) if Res_i^β is undefined and $(k_i - 1, \eta)$, where $\text{Res}_i^\beta = \mathfrak{C}_{k_i-1}(\mathcal{N}_\eta)$, otherwise. Then $\psi\colon \mathcal{J}^{Q^*_{\alpha+1}}_{\sigma\circ\pi_\beta(\lambda)} \to \text{Res}^\beta$ and $\sigma^\beta = \psi \circ (\sigma \restriction \mathcal{J}^{Q_\beta}_{\pi_\beta(\lambda)})$.

CLAIM 3. $\psi \restriction (\sup(\sigma \circ \pi_\beta{}''\kappa_{i-1})) = \text{identity}$.

PROOF. Suppose first that Res_i^β exists, so that $i \leq e$ and $\sigma = \sigma_i^\beta$. From claim 2 and the fact that π_β is a weak n-embedding we see that $\pi_\beta(\kappa_{i-1})$ is the projectum associated to the $(i-1)$st element of the $(n, \pi_\beta(\lambda))$-dropdown sequence of Q_β. As we remarked earlier, ψ is therefore the identity on $\sup(\sigma_i^\beta{}''\pi_\beta(\kappa_{i-1}))$, and this implies the claim.

Suppose next that Res_i^β is undefined, so that either $i = e+1$ or else $i = e$ and (c) of claim 2 holds with $u = \varnothing$. In either case the projectum associated to the last term of the $(n, \pi_\beta(\lambda))$ dropdown sequence of Q_β is at least $\sup(\pi_\beta{}''\kappa_{i-1})$. Thus $\sigma^\beta \restriction \sup(\pi_\beta{}''\kappa_{i-1})$ is the identity, but $\psi = \sigma^\beta$ and σ is the identity, so this implies the claim. □

We can now define $Q_{\alpha+1} = i^{\mathcal{U}}_{\beta,\alpha+1}(Q^*_{\alpha+1})$. Before we define $\pi_{\alpha+1}$ and verify the induction hypotheses, however, we must describe the agreement between $Q^*_{\alpha+1}$ and Res^α. Set

$$\gamma = \begin{cases} (\mu^+)^{\mathcal{P}^*_{\alpha+1}} & \text{if } \mathcal{P}^*_{\alpha+1} \models \mu^+ \text{ exists} \\ \text{OR}^{\mathcal{P}^*_{\alpha+1}} & \text{otherwise.} \end{cases}$$

CLAIM 4. $\gamma \leq \lambda = \text{lh}(E^{\mathcal{T}}_\beta)$, and if $\gamma = \text{OR}^{\mathcal{P}^*_{\alpha+1}}$ then $\mathcal{P}^*_{\alpha+1} = \mathcal{J}^{\mathcal{P}_\beta}_\lambda$ and $\mathcal{P}^*_{\alpha+1}$ is type II.

PROOF. If $\beta = \alpha$, then $(\mu^+)^{\mathcal{J}^{\mathcal{P}_\alpha}_\lambda}$ exists and $\mathcal{P}^*_{\alpha+1}$ is the shortest initial segment of $\mathcal{P}_\alpha$ over which a subset of μ not in $\mathcal{J}^{\mathcal{P}_\alpha}_\lambda$ is definable. Thus $(\mu^+)^{\mathcal{P}^*_{\alpha+1}} = (\mu^+)^{\mathcal{J}^{\mathcal{P}_\alpha}_\lambda} < \lambda \leq \text{OR}^{\mathcal{P}^*_{\alpha+1}}$, so $\gamma < \lambda \leq \text{OR}^{\mathcal{P}^*_{\alpha+1}}$.

If $\beta < \alpha$ then the subsets of μ in $\mathcal{P}_\alpha$ are just those in $\mathcal{J}^{\mathcal{P}_\beta}_\lambda$ and $\mathcal{P}^*_{\alpha+1}$ is the shortest initial segment of $\mathcal{P}_\beta$ over which a subset of μ not in $\mathcal{J}^{\mathcal{P}_\beta}_\lambda$ is definable, so if $(\mu^+)^{\mathcal{J}^{\mathcal{P}_\beta}_\lambda}$ exists then $(\mu^+)^{\mathcal{P}^*_{\alpha+1}} = (\mu^+)^{\mathcal{J}^{\mathcal{P}_\beta}_\lambda} < \lambda$. Otherwise μ is the largest cardinal of $\mathcal{J}^{\mathcal{P}_\beta}_\lambda$, so $\mathcal{P}^*_{\alpha+1} = \mathcal{J}^{\mathcal{P}_\beta}_\lambda$ since λ is definably collapsed over the active ppm $\mathcal{J}^{\mathcal{P}_\beta}_\lambda$. In this case we see also that $\mathcal{P}^*_{\alpha+1}$ is type II, since otherwise $\mu < \nu_\beta < \lambda$ and ν_β is a cardinal of $\mathcal{J}^{\mathcal{P}_\beta}_\lambda$. □

Claim 4 implies $\gamma \leq \kappa_{i-1}$. If $\kappa_{i-1} = \lambda$ then this is obvious. Otherwise κ_{i-1} is a cardinal of $\mathcal{J}^{\mathcal{P}_\beta}_\lambda$, since it is a projectum of some $\mathcal{J}^{\mathcal{P}_\beta}_\eta$ with $\eta \geq \lambda$. Since $\mu < \kappa_{i-1}$ by the choice of i, we have $\gamma \leq \kappa_{i-1}$.

The next claim shows that Res^α and $Q^*_{\alpha+1}$ have the agreement required for the use of the shift lemma.

CLAIM 5. (a) Res^α agrees with $Q^*_{\alpha+1}$ below $\sup(\sigma \circ \pi_\beta{}''\gamma)$.

(b) $\sigma^\alpha \circ \pi_\alpha \restriction \gamma = \sigma \circ \pi_\beta \restriction \gamma$.

PROOF. The proof of claim 5 is divided up into three subclaims.

Subclaim A. $Q^*_{\alpha+1}$ and Res^β agree below $\sup(\sigma \circ \pi_\beta{}''\gamma)$, and $\sigma \circ \pi_\beta \restriction \gamma = \psi \circ \sigma \circ \pi_\beta \restriction \gamma$.

This follows at once from claim 3 and the fact that $\gamma \leq \kappa_{i-1}$.

Subclaim B. If $\beta < \alpha$ then Res^β and Q_α agree below $\sup(\sigma \circ \pi_\beta{}''\gamma)$, and $\psi \circ \sigma \circ \pi_\beta \restriction \gamma = \pi_\alpha \restriction \gamma$.

Recall that $\psi \circ \sigma \circ \pi_\beta = \sigma^\beta \circ \pi_\beta$. This subclaim is therefore just our induction hypotheses on agreement. If Res^β is type I or type III then claim 4 yields $\gamma \leq \nu_\beta$ and we can apply H3. If Res^β is type II then $\gamma \leq \mathrm{lh}\, E^{\mathcal{T}}_\beta$ by claim 4 so we can apply H4.

Subclaim C. If $\beta < \alpha$ then Q_α and Res^α agree below $\sup(\sigma \circ \pi_\beta{}''\gamma)$ and $\pi_\alpha \restriction \gamma = \sigma^\alpha \circ \pi_\alpha \restriction \gamma$.

We have $\gamma \leq \lambda$, and $\sigma \circ \pi_\beta = \pi_\alpha \restriction \gamma$, so $\sup(\sigma \circ \pi_\beta{}''\gamma) \leq \pi_\alpha(\lambda)$. By claim 1, Q_α and Res^α agree below $\pi_\alpha(\lambda)$ and σ^α is the identity there.

Together, subclaims A, B and C yield claim 5. □

Now define, for $a \in [\nu_\alpha]^{<\omega}$ and appropriate f

$$\pi_{\alpha+1}\left([a, f]^{\mathcal{P}^*_{\alpha+1}}_{E^{\mathcal{T}}_\alpha}\right) = [\sigma^\alpha \circ \pi_\alpha(a), \sigma \circ \pi_\beta(f)]^{R_\beta}_{F^*}.$$

If $f = f_{\tau,q}$ then by "$\sigma \circ \pi_\beta(f)$" we mean $f_{\tau,\sigma\circ\pi_\beta(q)}$, the later function being defined over the ppm $Q^*_{\alpha+1}$. In order to see that $\pi_{\alpha+1}$ has the desired properties, it is useful to factor it. Let $k = \deg^{\mathcal{T}}(\alpha+1)$ and $Q'_{\alpha+1} = \mathrm{Ult}(Q^*_{\alpha+1}, F)$. Let $i\colon Q^*_{\alpha+1} \to Q'_{\alpha+1}$ be the canonical embedding and let $\pi'_{\alpha+1}\colon \mathcal{P}_{\alpha+1} \to Q'_{\alpha+1}$ be the weak k-embedding given by the shift lemma. Finally, let $\tau\colon Q'_{\alpha+1} \to Q_{\alpha+1}$ be the natural map given by $\tau\left([a, f]^{Q^*_{\alpha+1}}_{\sigma^\alpha\circ\pi_\alpha(E^{\mathcal{T}}_\alpha)}\right) = [a, f]^{R_\beta}_{F^*}$. Then $\pi_{\alpha+1} = \tau \circ \pi'_{\alpha+1}$ and we have the commutative diagram

$$\begin{array}{ccccc}
\mathcal{P}_{\alpha+1} & \xrightarrow{\pi'_{\alpha+1}} & Q'_{\alpha+1} & \xrightarrow{\tau} & Q_{\alpha+1} \\
\uparrow (i^*_{\alpha+1})^{\mathcal{T}} & & \uparrow i & \nearrow i^{\mathcal{U}}_{\beta,\alpha+1} & \\
\mathcal{P}^*_{\alpha+1} & \xrightarrow{\sigma\circ\pi_\beta} & Q^*_{\alpha+1} & &
\end{array}$$

In order to verify H1 we need to show that $\pi_{\alpha+1}$ is a weak k-embedding, where $k = \deg(\alpha+1)$, which means that we have to find a witness set X on which

$\pi_{\alpha+1}$ is $r\Sigma_{k+1}$ elementary. If $k = \deg(\alpha+1) = n$ and $\mathcal{P}^*_{\alpha+1} = \mathcal{P}_\beta$ then we can take the witnessing set to be $X = i^*_{\alpha+1}{}''X_\beta$, where X_β is a set witnessing that π_β is a weak k-embedding. Otherwise take $X = i^*_{\alpha+1}{}''\left|\mathcal{P}^*_{\alpha+1}\right|$. In either case the shift lemma implies that $\pi'_{\alpha+1}$ is $r\Sigma_{k+1}$ elementary on parameters from X. On the other hand the Łos theorem 4.1 implies that τ is $r\Sigma_{k+1}$ elementary on parameters from $i''\left|Q^*_{\alpha+1}\right|$, and since $\pi'_{\alpha+1}{}''X \subset i''\left|Q^*_{\alpha+1}\right|$ it follows that $\pi_{\alpha+1}$ is $r\Sigma_{k+1}$ elementary on parameters from X. Thus X witnesses that $\pi_{\alpha+1}$ is a weak k-embedding and we have verified H1. Induction hypothesis H2 comes from the commutativity of the diagram above.

We now verify H3 and H4. Let $\eta < \alpha+1$. If Res^η is type I or III then we must show that $Q_{\alpha+1}$ agrees with Res^η below ν^{Res^η} and moreover that $\pi_{\alpha+1} \restriction \nu_\eta = \sigma^\eta \circ \pi_\eta \restriction \nu_\eta$ and $\pi_{\alpha+1}(\nu_\eta) \geq \nu^{\mathrm{Res}^\eta}$. If Res^η is of type II, on the other hand, then we must show that $Q_{\alpha+1}$ agrees with Res^η below $\mathrm{OR}^{\mathrm{Res}^\eta}$ and moreover that $\pi_{\alpha+1} \restriction \mathrm{lh}\, E^{\mathcal{T}}_\eta = \sigma^\eta \circ \pi_\eta \restriction \mathrm{lh}\, E^{\mathcal{T}}_\eta$ and $\pi_{\alpha+1}(\mathrm{lh}\, E^{\mathcal{T}}_\eta) \geq \mathrm{OR}^{\mathrm{Res}^\eta}$.

We consider first the case $\eta = \alpha$. Set $\mu' = \pi_\beta(\mu)$. By claim 3, $\mathcal{J}^{Q^*_{\alpha+1}}_{\mu'} = \mathcal{J}^{\mathrm{Res}^\alpha}_{\mu'}$ so that

$$\mathcal{J}^{Q_{\alpha+1}}_{i^{\mathcal{U}}_{\beta,\alpha+1}(\mu')} = \mathrm{Ult}(\mathcal{J}^{Q^*_{\alpha+1}}_{\mu'}, F^*) = \mathrm{Ult}(\mathcal{J}^{\mathrm{Res}^\alpha}_{\mu'}, F^*),$$

where the ultrapowers are computed using all functions which are members of R_β, or equivalently of R_α, and which map $[\mu']^i$ into $\mathcal{J}^{Q^*_{\alpha+1}}_{\mu'}$ for some integer i.

Now the canonical embedding

$$\psi : \mathrm{Ult}_0(\mathrm{Res}^\alpha, F) \to \mathrm{Ult}(\mathrm{Res}^\alpha, F^*)$$

(where the first ultrapower uses all functions belonging to Res^α, and the second uses all functions in R_α) has critical point $\geq \nu^{\mathrm{Res}^\alpha}$ if Res^α is type I or III, and $\geq \mathrm{OR}^{\mathrm{Res}^\alpha} = \mathrm{lh}\, F$ if Res^α is type II. Moreover, $\mathrm{Ult}_0(\mathrm{Res}^\alpha, F)$ agrees with Res^α below $\mathrm{lh}\, F = \mathrm{OR}^{\mathrm{Res}^\alpha}$. As $i^{\mathcal{U}}_{\beta,\alpha+1}(\mu') > \mathrm{lh}\, F$, $Q_{\alpha+1}$ agrees with Res^α below $\nu^{\mathrm{Res}^\alpha}$ in the type I or III case, and below $\mathrm{OR}^{\mathrm{Res}^\alpha}$ in the type II case.

Next we consider the agreement of embeddings. Suppose first Res^α is type I or III, and $\xi < \nu_\alpha$. Then $\xi = [\{\xi\}, \mathrm{id}]^{\mathcal{P}^*_{\alpha+1}}_{E^{\mathcal{T}}_\alpha}$, where id = identity function, so

$$\pi_{\alpha+1}(\eta) = \left[\{\sigma^\alpha \circ \pi_\alpha(\xi)\}, \mathrm{id}\right]^{R_\beta}_{F^*} = \sigma^\alpha \circ \pi_\alpha(\xi)$$

as desired. Also, let $f \in |\mathcal{P}_\alpha| \cap |\mathcal{P}^*_{\alpha+1}|$ and $a \in [\nu_\alpha]^{<\omega}$ be such that $\nu_\alpha = [a,f]^{\mathcal{P}_\alpha}_{E^{\mathcal{T}}_\alpha} = [a,f]^{\mathcal{P}^*_{\alpha+1}}_{E^{\mathcal{T}}_\alpha}$. Then

$$\begin{aligned}\pi_{\alpha+1}(\nu_\alpha) &= [\sigma^\alpha \circ \pi_\alpha(a),\ \pi_\beta(f)]^{R_\beta}_{F^*} \\ &= [\sigma^\alpha \circ \pi_\alpha(a),\ \sigma^\alpha \circ \pi_\alpha(f)]^{R_\alpha}_{F^*} \\ &\geq [\sigma^\alpha \circ \pi_\alpha(a),\ \sigma^\alpha \circ \pi_\alpha(f)]^{\mathrm{Res}^\alpha}_{F}.\end{aligned}$$

But for $(E^{\mathcal{T}}_\alpha)_{a\cup\{\nu_\alpha\}}$ a.e. $(\bar{u}, v)$, $f(\bar{u}) = v$. Also $\sigma^\alpha \circ \pi_\alpha(\nu_\alpha) = \nu^{Res^\alpha}$, so $\sigma^\alpha \circ \pi_\alpha(f)(\bar{u}) = v$ for $(F)_{\sigma^\alpha\circ\pi_\alpha(a)\cup\{\nu^{\mathrm{Res}^\alpha}\}}$ a.e. $(\bar{u}, v)$. Thus

$$\nu^{\mathrm{Res}^\alpha} = [\sigma^\alpha \circ \pi_\alpha(a), \sigma^\alpha \circ \pi_\alpha(f)]^{\mathrm{Res}^\alpha}_F$$

and $\pi_{\alpha+1}(\nu_\alpha) \geq \nu^{\mathrm{Res}^\alpha}$, as desired.

These calculations carry over easily to the case Res^α is type II to give the agreement of embeddings facts in part (b) of the claim. We omit further detail.

We must now consider the case $\eta < \alpha$. Let's just prove (a), the proof of (b) being similar. So assume Res^η is type I or III.

From the $\eta = \alpha$ case we know that $Q_{\alpha+1}$ agrees with Res^α below $\nu^{\mathrm{Res}^\alpha}$. But we showed in the proof of claim 5 that Res^α agrees with Q_α below $\pi_\alpha(\mathrm{lh}\, E^{\mathcal{T}}_\eta)$. Also, $\pi_\alpha(\mathrm{lh}\, E^{\mathcal{T}}_\eta)$ is a cardinal of Res^α, hence $\pi_\alpha(\mathrm{lh}\, E^{\mathcal{T}}_\eta) \leq \nu^{\mathrm{Res}^\alpha}$. Thus $Q_{\alpha+1}$ agrees with Q_α below $\pi_\alpha(\mathrm{lh}\, E^{\mathcal{T}}_\eta)$. But by induction hypothesis, Q_α agrees with Res^η below ν^{Res^η}, and $\pi_\alpha(\nu_\eta) \geq \nu^{\mathrm{Res}^\eta}$. Thus $Q_{\alpha+1}$ agrees with Res^η below ν^{Res^η}, as desired. For agreement of embeddings, we argue similarly that $\pi_{\alpha+1} \restriction \nu_\alpha = \sigma^\alpha \circ \pi_\alpha \restriction \nu_\alpha$. Furthermore since $\mathrm{lh}\, E^{\mathcal{T}}_\eta$ is a cardinal of $\mathcal{P}_\alpha$ and $\mathrm{lh}\, E^{\mathcal{T}}_\eta < \mathrm{lh}\, E^{\mathcal{T}}_\alpha$, we know that $\mathrm{lh}\, E^{\mathcal{T}}_\eta \leq \nu_\alpha$, and since σ^α is the identity on $\pi_\alpha(\mathrm{lh}\, E^{\mathcal{T}}_\eta)$ we get that $\pi_{\alpha+1} \restriction \mathrm{lh}\, E^{\mathcal{T}}_\eta = \pi_\alpha \restriction \mathrm{lh}\, E^{\mathcal{T}}_\eta$. But then since $\pi_\alpha \restriction \nu_\eta = \sigma^\eta \circ \pi_\eta \restriction \nu_\eta$ by the induction hypothesis, $\pi_{\alpha+1} \restriction \nu_\eta = \sigma^\eta \circ \pi_\eta \restriction \nu_\eta$, as desired. Notice also that we get $\pi_{\alpha+1}(\nu_\eta) = \pi_\alpha(\nu_\eta) > v^{\mathrm{Res}^\eta}$ by induction.

This verifies H3 and H4. A much simpler coarse structural argument along the same lines gives H5. Finally, H6 is easy to check and H7 is vacuous in the successor case.

Now let λ be a limit ordinal with $\lambda < \theta = \mathrm{lh}\,\mathcal{T}$. We are given sequences $\mathcal{U} \restriction \lambda$, $\langle Q_\alpha \mid \alpha < \lambda\rangle$, and $\langle \pi_\alpha \mid \alpha < \lambda\rangle$ satisfying our inductive hypothesis, and must define $\mathcal{U} \restriction \lambda + 1$, Q_λ, and π_λ.

Let $c = [0, \lambda)_T = \{\alpha \mid \alpha T \lambda\}$. We claim that $\lim_{\alpha\in c} R_\alpha$ is wellfounded, where the limit is taken along the maps $i^{\mathcal{U}}_{\alpha\beta}$ for $\alpha, \beta \in c$.

For this it suffices, using results of [MS] asserting that $\mathcal{T}$ has at least one well founded branch, to show that if b is a branch of $\mathcal{T} \restriction \lambda$ which is cofinal in λ, and $b \neq c$, then $\lim_{\alpha\in b} R_\alpha$ is illfounded. So let b be such a branch.

We may assume $i^{\mathcal{U}}_{\alpha\beta}(Q_\alpha) = Q_\beta$ for all sufficiently large α and β in b, $\alpha < \beta$, as otherwise our last induction hypothesis 6(a) implies that $i^{\mathcal{U}}_{0b}(<_{\mathbb{C}})$ is illfounded, so $\lim_{\alpha\in b} R_\alpha$ is illfounded. (Here $i^{\mathcal{U}}_{\alpha b}$ is the canonical embedding from R_α, $\alpha \in b$, into $\lim_{\alpha\in b} R_\alpha$.) This in turn implies $D^{\mathcal{T}} \cap b$ is finite via 6(b).

Let $\mathcal{P}_b = \lim_{\alpha\in b} \mathcal{P}_\alpha$, and $Q_b = \lim_{\alpha\in b} Q_\alpha$, which is the common value of $i^{\mathcal{U}}_{\alpha b}(Q_\alpha)$ for $\alpha \in b$ sufficiently large. Then $\mathcal{P}_b$ exists as $D^{\mathcal{T}} \cap b$ is finite, and $\mathcal{P}_b$ is illfounded as $\mathcal{T} \restriction \lambda$ is simple and $b \neq c$. There is a natural $\pi : \mathcal{P}_b \to Q_b$ given by our commutativity hypothesis: $\pi(i^{\mathcal{T}}_{\alpha b}(x)) = i^{\mathcal{U}}_{\alpha,b}(\pi_\alpha(x))$, for $\alpha \in b$ sufficiently large. Thus Q_b is illfounded, and hence $\mathcal{R}_b$ is illfounded since $\mathcal{R}_b = \lim_{\alpha \in b} R_\alpha \models$ "Q_b is wellfounded".

We set $R_\lambda = \lim_{\alpha \in c} R_\alpha$, and this gives us $\mathcal{U} \restriction \lambda + 1$. Notice that $i^{\mathcal{U}}_{\alpha\lambda}(Q_\alpha)$ is constant on all sufficiently large $\alpha T \lambda$, as otherwise $i^{\mathcal{U}}_{0\lambda}(<_{\mathrm{C}})$ is illfounded. Set Q_λ equal to the eventual value of $i^{\mathcal{U}}_{\alpha\lambda}(Q_\alpha)$ for sufficiently large $\alpha T \lambda$. Set

$$\pi_\lambda(i^{\mathcal{T}}_{\alpha\lambda}(x)) = i^{\mathcal{U}}_{\alpha\lambda}(\pi_\alpha(x))$$

for $\alpha < \lambda$ sufficiently large, $\alpha T \lambda$.

Let $n = \deg^{\mathcal{T}}(\lambda) = \deg^{\mathcal{T}}(\alpha)$ for $\alpha T \lambda$ sufficiently large. It is easy to check that π_λ is a weak n-embedding which is $r\Sigma_{n+1}$ elementary on the appropriate set, and that π_λ commutes properly. Our last induction hypothesis is just the definition of Q_λ so we need only check that Q_λ and π_λ agree properly with Res^β and $\sigma^\beta \circ \pi_\beta$ for $\beta < \lambda$.

Let $\beta < \lambda$. We have already shown that if $\gamma > \beta$, then $\nu^{\mathrm{Res}^\gamma} > \nu^{\mathrm{Res}^\beta}$. But $\nu^{\mathrm{Res}^\gamma} \leq \mathrm{lh}\, E^{\mathcal{U}}_\gamma$, and thus $\beta < \gamma \Rightarrow \nu^{\mathrm{Res}^\beta} < i^{\mathcal{U}}_{\eta,\gamma+1}(\mathrm{crit}\ E^{\mathcal{U}}_\gamma)$ where $\eta = T\text{-pred}\,(\gamma + 1)$. As R_λ is wellfounded, we must have $\nu^{\mathrm{Res}^\beta} < \mathrm{crit}\ E^{\mathcal{U}}_\gamma$, for all sufficiently large $\gamma + 1\ T\ \lambda$. We can then find $\gamma + 1\ T\ \lambda$ sufficiently large that $\nu^{\mathrm{Res}^\beta} < \mathrm{crit}\ i^{\mathcal{U}}_{\gamma+1,\lambda}$ and $i^{\mathcal{U}}_{\gamma+1,\lambda}(Q_{\gamma+1}) = Q_\lambda$. By induction, $Q_{\gamma+1}$ agrees with Res^β below ν^{Res^β}. Q_λ agrees with $Q_{\gamma+1}$ below $\mathrm{crit}\ i^{\mathcal{U}}_{\gamma+1,\lambda}$. So Q_λ agrees with Res^β below ν^{Res^β}. For the embeddings, notice that $\beta < \gamma \Rightarrow \nu_\beta < \nu_\gamma < i^{\mathcal{T}}_{\eta,\gamma+1}(\mathrm{crit}\ E^{\mathcal{T}}_\gamma)$, where $\eta = T\ \mathrm{pred}(\gamma + 1)$. So we can assume the ordinal $\gamma + 1$ of the last paragraph is such that $i^{\mathcal{T}}_{\gamma+1,\lambda}$ is defined and $\nu_\beta < \mathrm{crit}\ i^{\mathcal{T}}_{\gamma+1,\lambda}$.

But then, for $\alpha < \nu_\beta$,

$$\pi_\lambda(\alpha) = \pi_\lambda(i^{\mathcal{T}}_{\gamma+1,\lambda}(\alpha)) = i^{\mathcal{U}}_{\gamma+1,\lambda}(\pi_{\gamma+1}(\alpha)) = \pi_{\gamma+1}(\alpha)\,.$$

Since $\pi_{\gamma+1} \restriction \nu_\beta = \sigma^\beta \circ \pi_\beta \restriction \nu_\beta$ by induction, $\pi_\lambda \restriction \nu_\beta = \sigma^\beta \circ \pi_\beta \restriction \nu_\beta$, as desired. This proves the agreement hypothesis in the case Res^β is type I or III. The type II case is almost the same.

We have completed the definition of $\mathcal{U} \restriction \theta = \mathcal{U}$. Assuming that θ is a limit ordinal, methods of [MS] yield a cofinal, wellfounded branch b of $\mathcal{U}$. It is easy to see (cf. the limit case above) that b is a wellfounded branch of $\mathcal{T}$. This was what we needed.

In the case θ is a successor, the fact that $\mathcal{U}$ can be extended freely one more step guarantees the same for $\mathcal{T}$, as desired.

The remaining clauses of k-iterability can be proved similarly, using that the corresponding operations on $L(V_\theta)$ yield wellfoundedness.

This completes the proof of 12.1. □

The 0-iterability of the bicephali and psuedo-premice arising in the construction of §11 can be proved similarly.

References

[D] A. Dodd, *Strong cardinals*, unpublished notes.

[DJ1] A. Dodd and R. B. Jensen, *The core model*, Annals of Math. Logic **20** (1981) 43-75.

[DJ2] A. Dodd and R. B. Jensen, *The Covering Lemma for K*, Annals of Math. Logic **22** (1982), 1-30.

[DJ3] The Covering Lemma for $L[U]$, Annals of Math. Logic **22** (1982) 127-155.

[DJ4] A. Dodd and R. B. Jensen, untitled notes on fine structure below a strong cardinal, unpublished.

[M74R] W. Mitchell, Sets constructible from sequences of measures: revisited, JSL **48** (1983), 600-609.

[M85] W. Mitchell, The core model for sequences of measures I, Math. Proc. of the Cambridge Phil. Soc. **95** (1984) 229-260.

[M?] W. Mitchell, The core model for sequences of measures II, to appear.

[MS] D. A. Martin and J. R. Steel, Iteration trees, to appear.

[MSS] W. Mitchell, E. Schimmerling, and J. Steel, The Covering Lemma up to One Woodin Cardinal, 1992, preprint.

[Sch] E. Schimmerling, Combinatorial Principles in the Core Model for One Woodin cardinal, 1992, preprint.o

[S?a] J. R. Steel, The Core Model Iterability Problem, in preparation.

[S?b] J. R. Steel, Projectively Wellordered Inner Models, 1993, submitted to Annals of Pure and Applied Logic.

[S?c] J. R. Steel, Inner Models with Many Woodin Cardinals, 1993, Annals of Pure and Applies Logic, to appear.

INDEX OF DEFINITIONS

Definitions not numbered in the text are indexed here by the number of the theorem, lemma, or definition immediately preceding; thus "1.0.3 ff." indicates an unnumbered definition occuring in the body of the text after definition 1.0.3.

Index